A Century of Mendelism in Human Genetics

T0186406

A Century of Mendelism in Human Genetics

Proceedings of a symposium organised by the Galton Institute
and held at the Royal Society of Medicine, London, 2001

EDITED BY

Milo Keynes, A. W. F. Edwards and Robert Peel

Galtonia *candicans*

The Galton Institute
19 Northfields Prospect
London, SW18 1PE

CRC Press
Taylor & Francis Group
Boca Raton London New York

CRC Press is an imprint of the
Taylor & Francis Group, an **informa** business

CRC Press
Taylor & Francis Group
6000 Broken Sound Parkway NW, Suite 300
Boca Raton, FL 33487-2742

First issued in paperback 2019

CRC Press is an imprint of Taylor & Francis Group, an Informa business

ISBN-13: 978-0-415-32960-6 (hbk)
ISBN-13: 978-0-367-39446-2 (pbk)
Library of Congress Card Number 2003065093

Library of Congress Cataloging-in-Publication Data

A century of Mendelism in human genetics / edited by Milo Keynes,
A.W.F. Edwards, and Robert Peel.
    p. cm.
  Includes bibliographical references and index.
  ISBN 0-415-32960-4 (alk. paper)
    1. Mendel's law—History. 2. Human genetics—History. I. Keynes, W.
Milo (William Milo) II. Edwards, A. W. F. (Anthony William Fairbank),
1935- III. Peel, Robert Anthony, 1953- IV. Title.
QH428.C46 2004
599.93'5—dc22   2003065093

Visit the Taylor & Francis Web site at
http://www.taylorandfrancis.com

and the CRC Press Web site at
http://www.crcpress.com

Contents

The First Fifty Years of Mendelism

Human Genetics from 1950

Appendix:

Notes on the Contributors

Sir Patrick Bateson, Sc.D., FRS, Professor of Ethology, University of Cambridge; Provost of King's College, Cambridge

William Bateson, FRS, Fellow, St John's College, Cambridge in 1900

John Bell, DM, FRCP, Regius Professor of Medicine, University of Oxford; John Radcliffe Hospital, Oxford

Michael Bulmer, D.Sc., FRS, Emeritus Professor, Biological Sciences, Rutgers University, New Jersey; The Old Vicarage, Chittlehampton, Umberleigh, Devon

Timothy M. Cox, M.D., FRCP, Professor of Medicine, University of Cambridge; Addenbrooke's Hospital, Cambridge

A.W.F. Edwards, Sc.D., Professor of Biometry, University of Cambridge; Gonville and Caius College, Cambridge

M.A. Ferguson-Smith, FRCPath, FRS, Emeritus Professor of Pathology, University of Cambridge; Department of Clinical Veterinary Medicine, Cambridge

Milo Keynes, M.D., FRCS, Galton Institute; Hon. Fellow, Darwin College, Cambridge

Alfred G. Knudson, M.D., Fox Chase Cancer Center, Philadelphia, Pennsylvania

Lucio Luzzatto, M.D., Director, Istituto Nazionale per la Ricerca sul Cancro, Genova

Eileen Magnello, D.Phil., Wellcome Trust Centre for the History of Medicine at University College London

Newton E. Morton, Hon. FRCP, Professor of Genetic Epidemiology, University of Southampton; Southampton General Hospital, Southampton

Sir David Weatherall, D.M., FRCP, FRS, Emeritus Regius Professor of Medicine, University of Oxford; Weatherall Institute of Molecular Medicine, John Radcliffe Hospital, Oxford

Preface

The symposium "A Century of Mendelism in Human Genetics", arranged by the Galton Institute and held at the Royal Society of Medicine in London on 30 and 31 October 2001, has a relevance to the Human Genome Project. Besides being of general medical concern, its proceedings will be of particular interest to departments of genetics and medical genetics, as well as to historians of science and medicine.

In 1901 William Bateson, FRS, Fellow of St John's College, Cambridge, published a lecture (reprinted here as an appendix), which he had delivered the year before to the Royal Horticultural Society in London. In this he recognised the importance of the work completed by Gregor Mendel in 1865 and brought it to the notice of the scientific world. Archibald Garrod, working on patients with alkaptonuria, read Bateson's paper and, realising the relevance of Mendel's law to human disease applied it to this "inborn error in metabolism" in 1902. He thus introduced Mendelism into what was to become medical genetics, as the term "genetics" was only coined by Bateson in 1905.

The contributions in the first part of the proceedings are historical. Francis Galton (1822-1911) began his efforts to discover the laws of inheritance in man on reading *On the Origin of Species* on its publication in 1859, writing his first work on heredity in 1865, which culminated in his "Theory of Ancestral Inheritance" in 1897. This theory was championed by the biometricians in bitter controversy with the Mendelians before the full acceptance of Mendelism. The second part is concerned with human genetics from 1950 and ends with a chapter on "Genetics and the Future of Medicine". The Galton Lecture for 2001 given by Allan Bradley, FRS, Director of the Sanger Institute, Hinxton, Cambridge, on "The Human Genome Project" has not been included.

There is no index to this book, as we found that making one was a work of supererogation, which we therefore abandoned. Half the entries, such as "Mendel" and "Galton", were giving so many leads to so many papers as to be unhelpful, and half were leading to a single paper already obviously relevant from its title alone.

A Note on the Galton Institute

This learned scientific society was founded in 1907 as the Eugenics Education Society, changing its name to the Eugenics Society in 1926, and becoming the Galton Institute in 1989. Francis Galton defined eugenics as "the scientific study of the biological and social factors which improve or impair the inborn qualities of human beings and of future generations" in 1883. The Institute is committed to environmental and genetic studies, and its membership is drawn from a wide range of disciplines, including the biological and social sciences, economics, medicine and law.

The First Fifty Years of Mendelism

1. The Introduction of Mendelism into Human Genetics

Milo Keynes

On 8 May 1900, William Bateson (1861-1926), Fellow of St John's College, Cambridge, gave a paper, "Problems of Heredity as a Subject for Horticultural Investigation", to the Royal Horticultural Society in London, published in the Society's journal[1] the next year (and here reprinted as an appendix). According to Robert Olby's reassessment of 1987[2], a few days before delivering the lecture Bateson was handed an offprint: "Sur la loi de disjonction des hybrides",[3] published in *Comptes Rendus de l'Académie des Sciences* sometime before 21 April. This had been sent to the hybridist Charles Hurst (1870-1947), a collaborator of Bateson since 1899, by its author, Hugo de Vries (1848-1935) of Amsterdam. Bateson's widow, Beatrice, mistakenly wrote in 1928[4] that Bateson "read Mendel's actual paper on peas for the first time" on the train to London and incorporated it in his lecture. In fact, he read de Vries's offprint, in which there was no reference to Mendel's paper. Bateson's lecture was then delivered without mention of Mendel's name or his work.

The *Comptes Rendus*[3] offprint was a summary of a paper[5] by de Vries on the segregation of hybrids published in the *Berichte der Deutschen Botanischen Gesellschaft* on 25 April, a copy of which de Vries sent to Hurst on 19 May and after Bateson's lecture. This did refer to the paper given in 1865 by Gregor Mendel (1822-1884) on "Versuche über Pflanzen-Hybriden"[6] ["Experiments in Plant Hybridisation"], which gave the results of Mendel's 20,000 experiments, made between 1857 and 1863, in crossing varieties of the garden pea, *Pisum sativum*. That paper was also the first to employ the theory of probability in biology. As shown by Olby,[2] Mendel had done the work to try and see if hybridisation gave a better explanation of the origin of species than transmutation, and not to search for a general theory of heredity.

On reading de Vries's *Berichte* paper,[5] Bateson, who was fluent in German, searched out Mendel's paper[6] and gave it a full citation in the printed text of his RHS paper[1] in 1901. In this he also made comment on de Vries's two papers,[3,5] but merely named the other publications[7,8,9,10,11] of 1900 by de Vries, Carl Correns (1864-1933) of Tübingen and Erich von Tschermak-Seysenegg (1871-1962) of Vienna, that also discussed Mendel's work. A translation of Mendel's paper appeared in the Royal Horticultural Society's journal[6] later in 1901, as well as being published with modifications in Bateson's *Mendel's Principles of Heredity: A Defence*, Cambridge, 1902. Soon after this, Bateson and Saunders[12] published the results of experiments on crossing poultry, and Lucien Cuénot (1866-1951)[13] on mice, each of which showed that Mendel's theory applied to the animal kingdom. Bateson's book *Mendel's Principles of Heredity*[14] (again with the translation of Mendel's paper) followed in 1909.

Bateson's recognition of the importance of Mendel's work has more significance than mere questions of priority in the publications of 1900. What genetics now has are two laws – the law of segregation and the law of independent assortment – both derived from Mendel, that summarise part of his paper. In fact, they were not named until long after his death.

Mendel's studies on hybridisation of the garden pea

The first use of *Pisum sativum* in the study of hybrids began in 1787. It was reported[15] to the Royal Society in 1799 by Thomas Andrew Knight (1759-1838) in experiments initially designed to see whether it was possible to confer characters on apples by artificial pollination. Knight crossed two varieties of peas and (to use today's Mendelian terms) discovered dominance in the first hybrid generation. He backcrossed the hybrids to the recessive parent and found both dominant and recessive types in the progeny. In 1822, John Goss[16] and Alexander Seton,[17] working independently (and with their results verified by Knight[18]), reported crossing different varieties of peas and discovering dominance in the first hybrid generation and the reappearance of both types in the second generation. Goss found three types in the third generation, the recessive, the heterozygous dominant (which produced some recessives as segregants) and the homozygous dominant that bred true. As noted by Conway Zirkle,[19] all of Mendelism had, in fact, been recorded except the independent inheritance of separate factors (itself described by Augustin Sageret[20] in melons in 1826), and a definite numerical ratio in the second generation (described by Johann Dzierzon[21] in bees in 1854).

Mendel chose *Pisum sativum* because its seed and plant have striking characteristics that are easily and reliably distinguishable, because it yields fertile hybrids and because the pollinated flower can easily be protected from cross-pollination. He crossed varieties differing from each other in one definite character and studied discontinuous variation of seven pairs of characters. He read his paper in two parts on 8 February and 8 March 1865 to 40 members of the Naturforschender Verein (Natural Science Society) of Brünn, Austria, now Brno, Czech Republic. When asked to publish the text, Mendel only handed it over after he had re-examined his "records for the various years of experimentation, and not having been able to find a source of errors"[22]. It was published in the Transactions of the Society[6] in 1866.

In its analysis of the inheritance of particular characters, Mendel's paper was entirely unlike all that had gone before. The aims of the hybridists and breeders were quite different. None of Mendel's audience were horticulturists or theoretical biologists, but, although the volume containing his work was only the fourth of a new publication, it was widely distributed by exchange arrangements with 115 universities, academies and scientific societies in Europe and the United States, so that every prominent biologist of the mid-nineteenth century had access to the paper[23]. The copy read by William Bateson was easily available to him from Cambridge University Library, as a copy possibly received on publication (and perhaps by exchange) was bound there in 1881.[2] However, though present as recently as 1985, it is now missing from the library.

Mendel ordered 40 offprints of his paper to distribute on New Year's Day of 1867, of which 8 have been traced.[22] Two are in Brno, and two others, one in Indiana University, Bloomington and one in Mishima in Japan, came from Brno in about 1921. One, now in Graz University, was sent to the botanist Franz Unger (1800-1870), who had been one of Mendel's teachers in Vienna (but was by then in retirement), and it remained unread. One went to the botanist Anton Kerner von Marilaun (1831-1898) at Innsbruck, who took a cynical view on laws of heredity and did not bother to read it, and one to another botanist, Carl Wilhem von Nägeli (1817-1870) in Munich, who lost no time in slitting open the pages, but whose behaviour towards the "amateur" Mendel was patronising and unhelpful.[23]

Nägeli suggested that Mendel should confirm his findings using hawkweed *Hieracium* hybrids, but this plant, which reproduces asexually, gave utterly disappointing results and led Mendel to doubt his original findings and to discontinue his botanical experiments by 1869.[24,25] Another reason for abandoning his work was that two years after the publication of the *Pisum* paper he had been elected abbot of the Augustinian monastery at Brünn and found he had little time for experimenting (besides complaining in a letter to Nägeli in 1867 of increasing girth limiting his botanical activity).[22]

Sometime before 1889, the eighth offprint, still in Amsterdam, reached the Dutch biologist Martinus Willem Beijerinck (1851-1931), who knowing Hugo de Vries was studying hybrids, sent it to him in 1900 at the time he was about to publish the results of his experiments.[26] De Vries's *Berichte* paper[5] (which mentioned the work of Mendel) was sent off to Berlin on 14 March and published on 25 April 1900.[2] De Vries followed this by sending a short summary[3] of it (albeit with no mention of Mendel) to *Comptes Rendus*, which was published in Paris before 21 April and got to Bateson by 7 May. This reached Carl Correns earlier than the paper published in Berlin, and as a result Correns[8] made the accusation that de Vries had been dishonest in delaying to make reference to Mendel in his writings.

De Vries, Correns and Tschermak studied different problems of plant hybridisation: each thought of himself as an innovator and each began to write the report on his experiments without knowing of Mendel's work. Even when they referred to it, they failed fully to understand it. Despite the wide impression given that all three rediscovered Mendel by independent search of the literature, both Correns and certainly Tschermak appear to have completed their reports after seeing de Vries's Berlin paper, where the reference to Mendel had been made from the reprint sent to de Vries by Beijerinck. No wonder that the identity of the rediscoverer of Mendel has been somewhat indecisive. It is, in any case, historically less important than the enthusiasm with which William Bateson hailed the significance of Mendel's work "with a kind of triumphant gladness".[4]

There were thirteen references to Mendel's *Pisum* paper in the literature between 1866 and 1900, most of them slight. They included mentions in the German botanical journal *Flora* in 1866, 1867 and 1872; the *Proceedings of the*

Viennese Academy of Science in 1871 and 1879; the thesis of C. A. Blomberg in 1872 for Stockholm University; the thesis of I. F. Schmalhausen for St Petersburg University in 1874; and the Royal Society's *Catalogue of scientific papers (1864-73)* issued in 1879; 4: 338.[22] More substantial is the reference in the publication[27] of 1869 by Hermann Hoffmann (1819-1891) on the determination of species and varieties, in which he attempted to refute Darwin's theory of evolution by denying the importance of variations as a basis for the formation of new species. Darwin himself made marginal notes in his copy of Hoffmann's book, but none against the pages in which there was mention of Mendel's hybridisation experiments. He also referred to Hoffmann's book in his *The Effects of Cross and Self-Fertilisation in the Vegetable Kingdom* (1876).

There is mention – but no signs of understanding – of Mendel's pea experiments in the book[28] of 1881 on plant hybrids by Wilhelm Olbers Focke (1834-1922), which was itself cited in two books[29,30] by L. H. Bailey (1858-1954) in 1892 and 1895. Charles Darwin (1809-1882) passed on his copy of Focke (received in November 1880) unread to G. J. Romanes (1848-1894), who then included Mendel's name in the list of hybridists at the end of the section on "Hybridism" in the *Encyclopaedia Britannica*[31] of 1881-1895 without any comment or apparently having read the pages in which reference is made to Mendel's experiments. Only in 1958 in his memoirs[32] did Tschermak state that in the winter of 1899-1900 he had found and used the reference to Mendel in Focke's book (thus belatedly reclaiming his priority in the rediscovery of Mendel in his paper[11] of 1900).

Darwin and Mendel

Mendel planned and began his experiments on *Pisum* two to three years before the publication of *On the Origin of Species* in 1859, before he could have heard of Darwin's theories. He knew no English and only bought and annotated the second German edition[33] of *On the Origin of Species* when it was published in 1863, though he must have heard of it earlier. Indeed, Alexander Makowsky (1833-1908) delivered a paper, "Ueber Darwin's Theorie der organischen Schöpfung" ["Theory of organic creation"]. on 11 January 1865 to the Brünn Natural Science Society (*Sitzungsberichte*, vol. IV 1865, pp. 10-18) a month before Mendel gave his first paper, though there is no evidence that Mendel was present at the meeting.[22]

Mendel visited England in 1862 as a member of a delegation to the Great Industrial Exhibition in London. However, he could not have gone to Down House and met Darwin that week as, after an attack of scarlet fever in the family, all the Darwins were away.[34] In any case, the Church authorities in Austria would scarcely have condoned an excursion to Downe, and any visit there by a Catholic priest would have caused much local comment.

Mendel bought most of Darwin's works, studying them closely and making frequent annotations. When he prepared his 1865 paper, Darwin's work was very much in his mind, and where he reflected upon it he did so objectively and without adverse criticism. He appears to have deliberately avoided opposing Darwin's views on inheritance by never mentioning his name in his lectures or scientific papers and only rarely in his drawn-out correspondence with Nägeli. He was not an adversary of Darwin's theories, but considered that an adequate theory of heredity was lacking from his system.[23,25] He clearly accepted the fact of evolution, and Sir Gavin de Beer[34] suggested in 1964 that he appears to have hoped that his discovery would provide something about evolution that was lacking, an explanation for the origin of a sufficient supply of heritable variation for natural selection to work on.

Darwin never visited Brünn, and his collection of offprints of scientific papers in Cambridge University Library does not include one of those sent out by Mendel in 1867. Two copies of the Transactions of the Natural Science Society of Brünn of 1866, received in 1867, were available to Darwin in London (at the Royal Society and the Linnean Society), but there is no evidence that he, or other Fellows for that matter, took them from the shelves.[23] A sentence in Hoffmann's book[27] dealing with Mendel:

> He believed that hybrids have the tendency to revert in later generations to the parental species

missed Mendel's important points, such as constant numerical ratios, of dominant and recessive characters and of non-blending hereditary transmission, and was hardly likely to have aroused Darwin's interest sufficiently for him to have consulted the original work.

The difficulty that Darwin had in reading scientific German could be another reason why he failed to perceive the importance of Mendel's laws of heredity for his theory of evolution, but his ignorance of mathematics is far less likely a reason, despite what he wrote in his autobiography[35]:

> I have deeply regretted that I did not proceed far enough at least to understand something of the great leading principles of mathematics; for men thus endowed seem to have an extra sense. But I do not believe that I should ever have succeeded beyond a very low grade.

In The Life and Letters of Charles Darwin[36], Francis Darwin (1848-1925) wrote of the great labour his father found in studying German texts and how little he could manage to read at a time. "He was especially indignant with Germans, because he was convinced that they could write simply if they chose ... He learnt German simply by hammering away with a dictionary; he would say that his only way was to read a sentence a great many times over, and at last the meaning occurred to him." Between 1860 and 1865 he paid his children's governess, Camilla Ludwig, to translate from the German for him,[37] and laughed "at her if she did not translate it fluently."[36]

Galton and Mendel

Before mentioning Mendel in his paper[1] of 1901, Bateson wrote that he expected that general expressions capable of wide application would be found that could justly be called "laws" of heredity, although, he added, there had so far been few investigations on the transmission of characters. Such laws had been obtained by statistical methods, and he acknowledged that the first systematic attempt to enunciate them had been due entirely to Francis Galton (1822-1911). Galton, half-first cousin of Charles Darwin (Erasmus Darwin

[1731-1802] was the grandfather of both), read *On the Origin of Species* on its appearance and immediately began to consider mankind's future in the light of the theory of evolution. His first work[38] on heredity appeared in 1865 and was followed by *Hereditary Genius*[39] in 1869 (Galton later wished he had used the word "talent" in the title to imply high ability rather than "genius"). In turn, came *A Theory of Heredity* [40] in 1875; *Typical Laws of Heredity*[41] in 1877; *Regression towards Mediocrity in Hereditary Stature*[42] in 1885; and *Natural Inheritance*[43] in 1889. Galton's "Theory of Ancestral Inheritance", derived from "The Law of Regression"[42] of 1885, appeared as "a new law of heredity"[44] in 1897.

His main effort from 1865 was to try to discover laws of inheritance in man. He rejected the idea of L. A. J. Quetelet (1796-1874) that all variation in human physical characteristics was an error about a type, and insisted "that the laws of Heredity were solely concerned with deviations expressed in statistical units". He saw that without variation there was no evolution. Deviation from the average was not an error. The answer, he thought, could be achieved by counting and figuring and by bringing quantitative methods into biology, with his maxim being "whenever you can, count".[45]

In 1877, encouraged by Darwin and with the backing of the botanist Joseph Hooker (1817-1911),[45] Galton began to breed the sweet pea, *Lathyrus odoratus*. He chose it because it had little tendency to cross-fertilise and all the peas in the pods were roughly the same size. He classified the seeds according to weight and gave a set of seven packets, each containing ten seeds of the same weight, to nine friends, including Darwin, to undertake their planting and culture – there is a letter from Downe in September 1877 advising him to "come down and sleep here and see them. They are grown to a tremendous height and will be very difficult to separate."[46]   With two failures, the plantings gave the produce of 490 carefully weighed seeds to create what was probably the first bivariate distribution. From this Galton constructed the first regression line (although his own term was "reversion").[41]

The data showed that seed weight was to some extent heritable and that quantitative traits are normally distributed in successive generations. John Edwards[47] has pointed out that, if Galton had been a better mathematician, his genetic law might have preceded Mendel's law in the scientific world's knowledge by twenty-three years. It was, however, anthropological evidence that Galton wanted and looked for by using pedigree analysis, twin studies and anthropometry. He cared "only for the seeds as means of throwing light on heredity in man".[48]

Bateson wrote to Galton on 8 August 1900, suggesting that he look up Mendel's paper "in case you may miss it. Mendel's work seems to me one of the most remarkable investigations yet made on heredity."[46] However, Galton, by then 78, failed to appreciate its significance and took little part in the controversy that then arose between the Mendelians and those who championed his law of ancestral heredity. Galton stated this law as follows: "The influence, pure and simple, of the mid-parent may be taken as a half, of the mid-grandparent as a quarter, of the mid-great-grandparent as an eighth, and so on."[42] It makes little sense under Mendelism, but follows naturally from Galton's theory of heredity, in which the hereditary particles are equally likely to be patent (expressed) or latent (not expressed). The ancestral model seemed to accommodate continuous variation satisfactorily and was taken up by the biometrician Karl Pearson (1857-1936), who developed the theory of multiple regression from it and generalised Galton's law as a prediction formula. Like his fellow biometrician, W. F. R. Weldon (1860-1906), Pearson did not accept Mendelism as the theory of inheritance.

In his paper[1] of 1901 to the Royal Horticultural Society, William Bateson summarised Galton's ancestral law and pointed out that it dealt with populations and with continuously varying characters. Mendel's laws, in contrast, applied to discontinuous variation in individuals. In publicising Mendelism, Bateson, Hurst and the other Mendelians became involved in a bitter argument with the biometricians, particularly over the inheritance of continuous characters, which did not appear to fit into any simple Mendelian pattern, but was later shown to be explicable in Mendelian terms by R. A. Fisher[49]. It took well over ten years for the arguments to fade and for Mendelian segregation to carry the day.

Archibald Garrod (1857-1936) first wrote on "an inborn error in metabolism" in his 1899 paper, "A contribution to the study of alkaptonuria"[50]. Garrod discussed this with Bateson on the publication of his 1901 RHS paper[1], resulting in Bateson reporting to the Evolution Committee of the Royal Society on 17 December 1901 (published[51] in 1902) on the significance of the excess of first cousin marriages amongst the parents of Garrod's patients with alkaptonuria, which, he said, gave "exactly the conditions most likely to enable a rare and usually recessive character to show itself."

Next year Garrod applied Mendel's law to the human in a further paper[52] on alkaptonuria: "It has recently been pointed out by Bateson that the law of heredity discovered by Mendel offers a reasonable account of such phenomena. … In the case of a rare recessive characteristic we may easily imagine that many generations may pass before the union of two recessive gametes takes place. … There seems to be little room for doubt … that a peculiarity of the gametes of both parents is necessary for its production." In 1908 Garrod gave his Croonian Lectures at the Royal College of Physicians on inborn errors of metabolism. In them he discussed albinism, alkaptonuria, cystinuria and pentosuria with a strong Mendelian flavour for each one.[53]

Thus Garrod may be considered to have been the first, in 1902, to introduce Mendelism into medical "genetics" (a term coined by Bateson in 1905 in a letter to the zoologist Adam Sedgwick [1854-1913] when they were looking for a term for the study of heredity and variation). The word "gene" first appeared, later still, in German in 1909 in a book[54] of twenty-five lectures by Wilhelm Johannsen (1857-1927), which were based on lectures given at the University of Copenhagen in 1903 and published in Danish in 1905. However, Galton had coined the word "Eugenic" from the Greek *eugenes* in 1883,[55] and the word "pangene" had been created by de Vries[56] for the bearers of the separate hereditary characters in 1889.

This historical introductory chapter to this book is followed by chapters covering the fifty years from when William Bateson first recognised the importance of Mendel's work and brought Mendelism to the notice of the scientific world in 1901. It was Archibald Garrod who, on reading Bateson's paper, saw the relevance of Mendel's laws to human disease, and in 1902 introduced Mendelism to what soon became, in fact, medical genetics. The remaining chapters are more clinical in discussing human genetics from 1950, with a final chapter examining genetics and the future of medicine.

Acknowledgement

The great help given by Peter Morgan, Cambridge University Medical Librarian, in writing this chapter is gratefully acknowledged.

References

1. W. Bateson. Problems of heredity as a subject for horticultural investigation. *J R Horticultural Soc* 1901; **25**: 54-61.
2. R. Olby. William Bateson's introduction of Mendelism into England: a reassessment. *Brit J History of Science* 1987; **20**: 399-420.
3. H. de Vries. Sur la loi de disjonction des hybrides. *Comptes Rendus de l'Académie des Sciences* 1900; **130**: 845-7.
4. B. Bateson. "Memoir" in *William Bateson. F.R.S., Naturalist. His essays and Addresses*. Cambridge, UK: Cambridge University Press, 1928, pp. 70, 73.
5. H. de Vries. Das Spaltungsgesetz der Bastarde. *Berichte der Deutschen Botanischen Gesellschaft* 1900; **18**: 83-90.
6. Gregor Mendel. Versuche über Pflanzen-Hybriden. *Verhandlungen des Naturforschenden Vereines in Brünn*. Abhandlungen IV Band 1865. Brünn: Im Verlage des Vereines 1866; **4**: 3-47. Eng. transl. by C. T. Druery: "Experiments in plant hybridization". *J R Hort Soc* 1901; **26**: 1-32.
7. H. de Vries. Sur les unités des caractères spécifiques et leur application á l'étude des hybrides. *Rev Génér Botanique* 1900; **12**: 257-71.
8. C. Correns. Mendels Regel über das Verhalten der Nachkommenschaft der Rassenbastarde. *Ber Deutsch Bot Gesell* 1900; **18**: 158-68.
9. C. Correns. Gregor Mendel. *Bot Zeit* 1900; **58**: 229-30.
10. C. Correns. Über Levkojenbastarde. Zur Kenntniss der Grenzen der Mendelschen Regeln. *Bot Centralblatt* 1900; **84**: 97-113.
11. E. Tschermak-Seysenegg. Über künstliche Kreuzung bei *Pisum sativum*. Zeitschr *f. d. landw. Versuchswesen in Österreich* 1900; **3**: 465-555.
12. W. Bateson, E.R. Saunders. Part II. Poultry. In *Reports to the Evolution Committee of the Royal Society*, London, 1902; **Report I**: 87-124.
13. L. Cuénot. La loi de Mendel et l'hérédité de la pigmentation chez les souris. *Arch Zool Exp Gén* 1902; 3e sér. **10**: 27-30.
14. W. Bateson. *Mendel's Principles of Heredity*. Cambridge UK: Cambridge University Press, 1909.
15. T.A. Knight. XII. An account of some experiments on the fecundation of vegetables. *Phil Trans R Soc London* 1799; **89**: 195-204.
16. J. Goss. On the variation in the colour of peas, occasioned by cross-impregnation. *Trans Hort Soc London* 1824; **5**: 234-5.

17. A. Seton. On the variations in the colours of peas from cross-impregnations. *Trans Hort Soc London* 1824; **5**: 236-7.

18. T.A. Knight. LVIII. Some remarks on the supposed influence of the pollen, in cross breeding, upon the colour of the seed-coats of plants, and the qualities of their fruits. *Trans Hort Soc London* 1824; **5**: 377-80.

19. C. Zirkle. Some oddities in the delayed discovery of Mendelism. *J Heredity* 1964; **55**: 65-72.

20. A. Sageret. Considérations sur la production des hybrides, des variantes et des variétés en général, et sur celles de la famille des Cucurbitacées en particulier. *Ann Sci Nat* 1826; **8**: 294-313.

21. J. Dzierzon. Die Drohnen. *Der Bienenfreund aus Schliesen* 1854; **8**: 63-4.

22. V. Orel. *Gregor Mendel – the First Geneticist*. London: Oxford University Press, 1996, pp. 69-73, 96, 188-99, 276-7.

23. E. Posner, J. Skutil. The great neglect: the fate of Mendel's classic paper between 1865 and 1900. *Med Hist* 1968; **12**: 122-36.

24. G. Mendel. Über einige aus künstlichen Befruchtung gewonnenen Hieracium-Bastarde. *Verh Naturf Ver Abh Brünn* 1870; **8**: 26-31.

25. E. Posner. The enigmatic Mendel. *Bull Hist Med* 1966; **40**: 430-40.

26. T. J. Stomps. On the discovery of Mendel's work by Hugo de Vries. *J Heredity* 1954; **45**: 283-4.

27. H. Hoffmann. *Untersuchungen zur Bestimmung des Werthes von Species und Varietät: ein Beitrag zur Kritik der Darwinschen Hypothese*. Giessen: J. Richter, 1869, p.136.

28. W.O. Focke. *Die Pflanzen-Mischlinge. Ein Beitrag zur Biologie der Gewächse*. Berlin: Gebrüder Bornträger, 1881.

29. L.H. Bailey. "Cross-breeding and hybridizing." In *The Rural Library Series*. New York: Rural Publishing Co., 1892; **1**: 1-44.

30. L.H. Bailey. *Plant Breeding*. New York: Macmillan, 1895.

31. G.J. Romanes. "Hybridism". In *Encyclopaedia Britannica*, 9th ed., 1881-95; **12**: 422-6.

32. E. Tschermak-Seysenegg. *Leben und Wirken eines österreichischen Pflanzenzüchters*. Berlin: Parey, 1958.

33. C. Darwin. *Über die Entstehung der Arten im Tier- und Pflanzenreiche durch natürliche Züchtung*. German transl. by H. G. Bronn of 3rd English ed. of 1861. 2nd ed. Stuttgart: Schmeizerbart, 1863.

34. G. de Beer. Mendel, Darwin, and Fisher. *Notes, Records R Soc London* 1964; **19**: 192-226.

35. N. Barlow (ed.). *The Autobiography of Charles Darwin 1809-1882*. London: Collins, 1958, p. 58.

36. F. Darwin (ed.). *The Life and Letters of Charles Darwin*. 3 vols. London: John Murray, 1887; **1**: 126.

37. F. Burkhardt, D. M. Porter, et al. (eds.). *The Correspondence of Charles Darwin 1864*. Cambridge, UK: Cambridge University Press, 2001; **12**: 367, notes pp. 368, 412.

38. F. Galton. Hereditary talent and character. *Macmillan's Magazine* 1865; **12**: 157-66, 318-27.

39. F. Galton. *Hereditary Genius*. London: Macmillan, 1869.

40. F. Galton. A theory of heredity. *Contemporary Review* 1875-6; **27**: 80-95.

41. F. Galton. Typical laws of heredity. *Nature* 1877; **15**: 492-5, 512-4, 532-3.

42. F. Galton. Presidential address, Section of Anthropology, *Brit Assoc Report* 1885; **55**: 1206-14, and *Nature* 1885; **32**: 507-10; Regression towards mediocrity in hereditary Stature. *J Anthropological Inst* 1885; **15**: 246-63.

43. F. Galton. *Natural Inheritance*. London: Macmillan, 1889.

44. F. Galton. The average contribution of each several ancestor to the total heritage of the offspring. *Proc Roy Soc* 1897; **61**: 401-13; A new law of heredity. *Nature* 1897; **56**: 235-7.

45. F. Galton. *Memories of My Life.* London: Methuen, 1908.

46. N.W. Gillham. *A Life of Sir Francis Galton.* London: Oxford University Press, 2001, pp. 202-5, 305.

47. J.H. Edwards. "Francis Galton - numeracy and innumeracy in genetics". In Milo Keynes (ed.). *Sir Francis Galton, FRS - the Legacy of His Ideas.* London: Macmillan, 1993, p. 86.

48. D.W. Forrest. *Francis Galton - the Life and Work of a Victorian Genius.* London: Paul Elek, 1974, p. 188.

49. R.A. Fisher. The correlation between relatives on the supposition of Mendelian inheritance. *Trans Roy Soc Edinburgh* 1918; **52**: 399-433.

50. A.E. Garrod. A contribution to the study of alkaptonuria. *Med-chir Trans* 1899; **82**: 369-94; *Proc Roy Med Chir Soc* 1899; **N.S. II**: 130-5.

51. W. Bateson, E.R. Saunders. Part III. The facts of heredity in the light of Mendel's discovery. In *Reports to the Evolution Committee of the Royal Society*, London, 1902; **Report I**: 125-60, footnote 133-4.

52. A.E. Garrod. The incidence of alkaptonuria: a study in chemical individuality. *Lancet* 1902; **ii**: 1616-20.

53. A.G. Bearn. *Archibald Garrod and the Individuality of Man.* Oxford: Clarendon Press, 1993, pp. 76-87.

54. W.L. Johannsen. *Elemente der Exakten Erblichkeitslehre.* Jena: G. Fischer, 1909, p. 24.

55. F. Galton. *Inquiries into Human Faculty and Its Development.* London: Macmillan, 1883, p. 24.

56. H. de Vries. *Intracellulare Pangenesis.* Jena: G. Fischer, 1889.

2. Galton's Theory of Ancestral Inheritance

Michael Bulmer

In 1908 William Bateson's close collaborator, Reginald Punnett, read a paper to the newly formed Royal Society of Medicine on "Mendelism in relation to disease." Dr. Vernon (Oxford) sent a written contribution to the discussion, pointing out that the three diseased conditions quoted by Punnett as examples of Mendelism were very rare, and that among normal characters only eye colour had been shown to conform to the law. He continued: "The vast amount of work done by Galton, Pearson and others on the transmission of ... blended characters and their relation to the characters of the parents, grandparents, &c., was practically ignored by the Mendelians. For the average medical man a knowledge of the laws of ancestral heredity ... appeared more important than a knowledge of the segregated transmission of a few very rare diseases, interesting as such cases were." Punnett remained firm in his reply: "Dr Vernon's letter raised the old controversy between the Mendelians and the biometricians, and dwelt upon the practical value of the law of ancestral heredity as defined by Pearson and others. But it did not seem to him that a law which utterly collapsed before such simple facts as the production of colour from two pure strains of poultry or sweet peas was likely to be of much value to the average medical man or to anybody else."

This exchange of views at this Society nearly a century ago illustrates the bitterness of the dispute between the ancestrians and the Mendelians and prompts us to enquire what the law of ancestral heredity was and why it was so passionately debated.

Galton's Formulation of the Ancestral Law

In the introduction to *Natural Inheritance* (1889) Galton stated the three main questions to be addressed. The second question led to the ancestral law:

> A second problem regards the average share contributed to the personal features of the offspring by each ancestor severally. Though one half of every child may be said to be derived from either parent, yet he may receive a heritage from a distant progenitor that neither of his parents possessed as *personal* characteristics. Therefore the child does not on the average receive so much as one half of his *personal* qualities from each parent, but something less than a half. The question I have to solve, in a reasonable and not merely in a statistical way, is, how much less?

This passage distinguishes between what we call the genotype, one half of which is derived from each parent, and the personal features or phenotype. Galton was asking how much of a child's phenotype was derived from each of its parents, how much from its grandparents, and so on. This question does not make sense under a Mendelian model, but it makes good sense under Galton's model of heredity.

In the 1870s Galton developed a theory of heredity based on Darwin's theory of pangenesis, but rejecting the transportation of the hereditary particles or gemmules in the body. He supposed that a few of the gemmules in the fertilised

13

ovum became patent and were developed into the cells of the adult person, with the residue remaining latent. He also supposed that the germ cells contributing to the next generation were derived from the latent residue of gemmules that had not developed into the adult person; he did not appreciate the difficulty of accounting for the correlation between parent and offspring under this model (Bulmer, 1999).

He subsequently came to suppose that latent and patent gemmules were *equally* frequent and had the *same* chance of being transmitted to the next generation. He briefly described this idea, under which the correlation between parent and offspring can be explained much more easily, in *Natural Inheritance*, though he did not explicitly acknowledge the change from his previous theory. Under this model it was natural for him to ask: How many of the patent particles in a particular individual were patent in a parent? How many were last patent in a grandparent? and so on. This led him to propose the law of ancestral heredity, which he stated in 1885 as follows: "The influence, pure and simple, of the mid-parent may be taken as $\frac{1}{2}$, of the mid-grandparent $\frac{1}{4}$, of the mid-great-grandparent $\frac{1}{8}$, and so on. That of the individual parent would therefore be $\frac{1}{4}$, of the individual grandparent $\frac{1}{16}$, of an individual in the next generation $\frac{1}{64}$, and so on." This follows immediately from his modified theory of heredity, though he did not derive the law in this way.

This is the statement of the ancestral law as a representation of the separate contributions of each ancestor, on average, to the expressed phenotype of the offspring. The law can also be interpreted as a prediction formula for predicting the offspring value y_0 given the values of the mid-parent y_1, of the mid-grandparent y_2, and of all the more remote mid-ancestors from the regression formula:

$$E(y_0 \mid y_1, y_2, y_3, \cdots) = \tfrac{1}{2} y_1 + \tfrac{1}{4} y_2 + \tfrac{1}{8} y_3 + \cdots \qquad [1]$$

Galton assumed that these two interpretations of the ancestral law, as a representation of the separate contributions of each ancestor, on average, to the expressed phenotype of the offspring and as a prediction formula for predicting the value of a trait from ancestral values, were equivalent, though this is only approximately true. Because the regression coefficients in eqn 1 sum to unity, the same law applies to the corresponding deviations from the mean, $d_i = y_i - \mu$.

Galton first derived the ancestral law in 1885 by an ingenious, semi-empirical argument. The argument had several errors due to his failure to understand multiple regression theory, but it was in principle substantially correct. The argument can be rephrased in modern terminology as follows. Suppose that the regression of an individual on all his ancestors is

$$d_0 = \beta_1 d_1 + \beta_2 d_2 + \beta_3 d_3 + \cdots + e \qquad [2]$$

where e is the prediction error, and the β's are regression coefficients to be determined. Taking Expected values conditional on the mid-parental deviation yields

$$E(d_0 \mid d_1) = \beta_1 d_1 + \beta_2 E(d_2 \mid d_1) + \beta_3 E(d_3 \mid d_1) + \cdots \qquad [3]$$

Galton had found that the regression of offspring on mid-parent was $2/3$, while the regression of mid-parent on offspring was $1/3$. The regression of mid-grandparent on mid-parent must also be $1/3$, and he argued by analogy that the regression of mid-great-grandparent on midparent was $1/9$, and so on. (This was his first mistake.) Hence,

$$\frac{2}{3} = \beta_1 + \frac{1}{3}\beta_2 + \frac{1}{9}\beta_3 + \cdots \qquad [4]$$

To evaluate the partial regression coefficients in eqn 4, Galton considered two limiting hypotheses. Under the constant hypothesis, $\beta_i = \beta$ for all i, so that $\beta = 4/9$. Under the geometric decrease hypothesis, $\beta_i = \beta^i$, so that $\beta = 6/11$. Galton now remarked that the two estimates of β were nearly the same, and that their average was nearly $\frac{1}{2}$, and he concluded that $\beta_1 = \frac{1}{2}$, $\beta_2 = \frac{1}{4}$, $\beta_3 = 1/8$, and so on, leading to the law of ancestral inheritance in eqn 1. This part of the argument is rather contrived.

When he returned to the subject in 1897, confirming the law from data on coat colour in basset hounds, he replaced this semi-empirical argument by two much less convincing *a priori* arguments. He first appealed to recent discoveries about the reduction division of the germ cells as lending plausibility to the ancestral law; he appears to have been misled by a false analogy between the halving of the number of chromosomes in the reduction division and the coefficients in the ancestral law. He then argued that it was plausible to assume the geometric relationship $\beta_i = \beta^i$ and that the terms must sum to unity. Hence, $\beta = \frac{1}{2}$, giving the ancestral law, since this is the only geometric series whose terms sum to unity.

Karl Pearson's Development of the Ancestral Law

To Galton the ancestral law had a dual interpretation, as a representation of the contributions of different ancestors and as a multiple regression formula for predicting the value of a trait from ancestral values. He assumed incorrectly that these two interpretations were equivalent. Karl Pearson, who was a strong mathematician, was stimulated by this work to develop the modern theory of multiple regression and to apply it to the ancestral law. Pearson had a phenomenalist philosophy of science, which he summarised in *The Grammar of Science* in the phrase: "All science is description and not explanation." He was thus attracted to a model-free statistical theory that would provide an economical description of the facts of heredity. He therefore regarded the ancestral law purely as a prediction formula and rejected the interpretation of ancestral contributions, writing in his *Life of Galton* (vol 3A): "The term 'contribution of an ancestor' should be interpreted as, or be replaced by, 'contribution of the ancestor to the prediction formula.' *It is in no sense a physical contribution to the germ-plasms on which the somatic characters of the offspring depend. ...* The fact, I think, is that Galton's own ideas at this time were obscured by his belief that the ancestors actually did contribute to the heritage."

Pearson developed the theory of multiple regression in 1896, showing how the coefficients in a multiple regression formula could be calculated from the pairwise correlation coefficients. In particular, he calculated the joint regression of offspring on the parents and more remote ancestors on Galton's geometric assumption that if the offspring-parent correlation is r, then the offspring-grandparent correlation will be r^2, and so on. He showed that, under this assumption, if the values of both parents are known, then the regression coefficients on the grandparents and any more remote ancestors are all zero, in contradiction of the ancestral law. In the following year (1897) Galton published his paper verifying the ancestral law as a formula for predicting coat colour in basset hounds. Pearson was convinced by this paper that the ancestral law was basically correct, and he concluded that Galton's assumption that the correlation coefficients form the geometric series $r_i = r^i$, with $r = 1/3$, must be wrong. In 1898 he set out to find what correlations of r_i would lead to the ancestral law. He found that the modified geometric law $r = 0.6(1/2)^i$ would lead to a multiple regression on mid-parent, mid-grandparent, and so on in which the partial regression coefficients are $1/2$, $1/4$, $1/8$, and so on. Pearson remarked that Galton's first estimate of the regression of offspring on mid-parent for height was 0.6, in exact agreement with the value of $2r_1$ for the regression of offspring on mid-parent, and that he afterwards changed it to $2/3$, which is in less good agreement.

However, he was worried that the inflexibility of the law made the strength of inheritance an absolute constant. He proposed that the law could be generalised by representing the β_i s by the relationship $\beta_I = \gamma\beta^i$, where γ measured the strength of inheritance. If the β_i s are constrained to sum to unity, this leaves one free parameter to be estimated from the data. To give this result, the ancestral correlations must be of the form $r_i = \alpha(1/2)^i$. With this modification, he concluded that "the law of ancestral heredity is likely to prove one of the most brilliant of Mr Galton's discoveries; it is highly probable that it is the simple descriptive statement which brings into a single focus all the complex lines of hereditary influence. If Darwinian evolution be natural selection combined with *heredity*, then the single statement which embraces the whole field of heredity must prove almost as epoch-making to the biologist as the law of gravitation to the astronomer."

The Ancestral Law after 1900

Pearson regarded the ancestral law as an empirical generalisation. After 1900 he adopted an ambivalent attitude towards Mendelism, suspicious of its universality but sufficiently interested to investigate its mathematical consequences. In 1904 he considered a Mendelian model with n diallelic loci with complete dominance and with equal gene frequencies. Coding the phenotypic value as the number of dominant loci, he showed that, under random mating, the regression of the child on a single parent is linear with a slope of $1/3$; but the regression of the offspring deviation on that of both parents has the non-linear form

$$E(d_O \mid d_F, d_M) = \tfrac{1}{3}(d_F + d_M) - \tfrac{4}{9n} d_F d_M \qquad [5]$$

though the non-linearity becomes negligible with a large number of loci. He concluded that Mendelian theory was "not sufficiently elastic to cover the observed facts," since it required a linear regression of $1/3$. To the end of his life he thought that the ancestral law made Mendelism redundant, and as late as

1930 he wrote: "It has often been suggested that the Ancestral Law is contradicted by the discoveries of Mendel and his fellows; it is needless to say that this cannot be the case, for the law does not depend on any mechanism of the germ plasma." This statement is disingenuous because it ignores the requirement in the ancestral law that the multiple regression is linear.

Pearson regarded the ancestral law as a purely descriptive, statistical law, but most of his ancestrian colleagues took at face value Galton's first interpretation of the law, as a representation of the average contibutions of different ancestors. To them dominance presented a more fundamental problem than a technical question about linearity of regression. If a yellow pea from a pure race is crossed to a green pea from a pure race, all the offspring will be yellow, but one quarter of the offspring in the F_2 generation will be green. Under Mendelism, these extracted green peas will breed true, despite their yellow grandparents, in direct contradiction of the idea of grandparental contributions to their grandchildren. In particular, Pearson's close colleague Weldon took up the fight against Mendelism, writing in 1902: "The fundamental mistake which vitiates all work based upon Mendel's method is the neglect of ancestry, and the attempt to regard the whole effect upon offspring, produced by a particular parent, as due to the existence in the parent of particular structural characters."

Attempts to give meaning to ancestral contributions in a literal sense were, of course, mistaken. The statistician G.U. Yule provided a modern interpretation of the ancestral law under Mendelism in two remarkable papers in 1902 and 1906. In 1902 he defined the law as a prediction formula: "This law then, that *the mean character of the offspring can be calculated with the more exactness, the more extensive our knowledge of the corresponding characters of the ancestry,* may be termed the Law of Ancestral Heredity." His main contribution was to provide an explanation of this fact. He argued that "the somatic character of an individual is not … an absolute guide to the character of the ovum from which he sprang nor, *a fortiori*, to the mean character of the germ cells which he produces." There were two reasons for this, environmental variability and dominance, both of which ensure that ancestors can contribute information about the genotype of the offspring. He concluded that "the law of ancestral heredity need not in any way imply actual physical contributions of the ancestry to the offspring. The ancestry of an individual may serve as guides to the most probable character of his offspring simply because they serve as indices to the character of his germplasm as distinct from his somatic characters."

Yule's second paper in 1906 was a response to Pearson's conclusion in 1904 that Mendelian theory was "not sufficiently elastic to cover the observed facts." Yule suggested that the theory could be made more elastic by dropping the requirement of complete dominance and by allowing for environmental variability. He gave an example which was later generalised to show that the parent-child correlation is $(\frac{1}{2})h^2$, where the heritability h^2 is defined in the usual way, and that the correlation between an individual and an ancestor i generations back is $r_i = (\frac{1}{2})^i h^2$. This is of the form that gives Pearson's generalisation of the ancestral law, provided that the multiple regression is linear, which is likely to be approximately true for polygenic characters.

References

Bulmer, M. 1999. The development of Francis Galton's ideas on the mechanism of heredity. *J. Hist. Biol.,* **32,** 263-292.

Galton, F. 1889. *Natural Inheritance.* London: Macmillan.

Galton, F. 1885. Regression towards mediocrity in hereditary stature. *J. Anthropol. Inst.,* **15,** 246-263.

Galton, F. 1897. The average contribution of each several ancestor to the total heritage of the offspring. *Proc. R. Soc.,* **61,** 401-413.

Pearson, K. 1896. Mathematical contributions to the mathematical theory of evolution. – III. Regression, heredity, and panmixia. *Phil. Trans. R. Soc.,* **187,** 253–318.

Pearson, K. 1898. Mathematical contributions to the mathematical theory of evolution. On the law of ancestral heredity. *Proc. R. Soc.,* **62,** 386-412.

Pearson, K. 1904. Mathematical contributions to the theory of evolution. – XII. On a generalised theory of alternative inheritance, with special reference to Mendel's laws. *Phil. Trans. Roy. Soc.* A, **203,** 53-86.

Pearson, K. 1930. *Life of Galton,* vol IIIA: "Correlation, Personal Identification and Eugenics". Cambridge University Press.

Pearson, K. 1930. On a new theory of progressive evolution. *Ann. Eugenics,* **4,** 1-40.

Pearson, K. 1937. *The Grammar of Science.* London: Dent.

Punnett, R.C. 1908. Mendelism in relation to disease. *Proc. R. Soc. Med.,* **1,** 135-168 (Epidemiological Section).

Weldon, W. F. R. 1902. Mendel's laws of alternative inheritance in peas. *Biometrika,* **1,** 228-254.

Yule, G.U. 1902. Mendel's laws and their probable relations to intra-racial heredity. *New Phytologist,* **1,** 193-207, 222-238.

Yule, G.U. 1906. On the theory of inheritance of quantitative compound characters on the basis of Mendel's laws – A preliminary note. *Report 3rd Int. Conf. Genet.,* 140-142.

3. The Reception of Mendelism by the Biometricians and the Early Mendelians (1899-1909)

Eileen Magnello

At the Galton Institute's 2000 conference, which was the first part of "A Century of Mendelism", the historian of science, Peter Bowler, examined the conventional views on the rediscovery of Mendelism in the early years of the twentieth century.[1] When Peter discussed the historiography of Gregor Mendel's role on producing a theory of particulate inheritance, through his experimental work on garden peas in the 1860s, he emphasised that

> the rediscovery of Mendelism cannot be understood as a simple recognition by three scientists independently that a particulate model of heredity self-evidently offered the basis for the complete reformulation of scientific thinking in this area.[2]

Until the 1970s, this conventional account of the rediscovery of Mendelism was attributed to Carl Correns, Hugo de Vries and Erich von Tschermak who had come across Mendel's paper in 1900. Historians of science who then began to re-examine Mendel's place in the development of genetics challenged this orthodox triumphalist account of the simultaneous "re-discovery" of Mendel's laws by these three scientists.[3]

These historians showed that Mendel's main interest was in the hybridisation of species as an alternative to evolution rather than in theories of inheritance. It was thus argued that "Mendel's revival in 1900 took place in the context of a priority dispute between Correns and de Vries … [which] led scientists to overlook the original intention of the earlier research".[4] Whilst the heroic account of Mendel has filled a central historiographical role for a number of geneticists throughout the twentieth century, when they attempted to explain why Mendel's work was not accepted immediately in the community, Karl Pearson has usually been portrayed as the anti-hero who delayed scientific progress. In order to consider Pearson's role in this debate, the reception of Mendelism in the early 1900s and the subsequent debates from the biometricians, including Pearson, W.F.R. Weldon, Francis Galton and George Udny Yule, and from the early Mendelians such as William Bateson and William Ernest Castle will be examined.

An energetic and enterprising polymath, Karl Pearson's interests ranged from astronomy, mechanics, meteorology and physics to the biological sciences. Having started his career as an elastician (that is, someone who devised mathematical equations for elastic properties of matter), Pearson pursued a number of areas before he settled on mathematical statistics. He studied philosophy in Germany in 1879, which he abandoned because "philosophy made him miserable", and then decided to study Roman Law in Berlin and was called to the Bar, but by 1880 he was "tired of the law", as he found it to be a rather depressing practice. Forsaking the law, he embraced the study of medieval German folklore and literature, and became so competent that he was short-listed for the newly created post in German in Cambridge in the summer of 1884. Despite this success, he "longed to be working with symbols and not words".

After having been rejected from six mathematical posts over a period of two years, he received the Chair of "Mechanism and Applied Mathematics" at University College London (UCL) in the summer of 1884. Pearson also played a pivotal role in the institutional development of UCL, as he created a Department of Structural Engineering (now Civil Engineering) in 1892, established a Department of Astronomy in 1904 and founded the Biometric School in 1893, which was incorporated into the Drapers Biometric Laboratory ten years later and became the Department of Applied Statistics in 1911. As I have argued elsewhere, Pearson's change of careers from mathematical physics to establishing biometry as a new discipline, which provided the foundation to the modern theory of statistics, was due largely to the influence of his closest friend and colleague, W.F.R. Weldon.[5]

Pearson's status as the anti-hero in this debate on the reception of Mendelsim is due largely to the long-standing claims, which were made for virtuallly the entire twentieth century, that he rejected Mendelism as a theory of inheritance. This is a view that was first expressed by William Bateson in 1902 and by William Ernest Castle in 1903.[6] Later generations of such biologists as J.B.S. Haldane, Lancelot Hogben, Julian Huxley, Reginald Punnett, Alfred Henry Sturtevant and Sewall Wright, all of whom were familiar with the works of Bateson and Castle, continued to perpetuate these views during their lifetimes.[7] Moreover, all of these biologists also claimed (somewhat Whiggishly) that "Pearson delayed scientific progress for more than twenty years". Subsequently, these views have shaped Pearson's historiography during the past thirty years to the extent that such historians of science as Joan Fisher Box, Bernard Norton, Robert Olby, R.G. Swinburne and William Provine have all been predisposed to accept Bateson's and Castle's claims that Pearson rejected or opposed Mendelism.[8]

In 1971, however, Peter Froggatt and N.C. Nevin cast doubt on these claims, but their views never became a part of the discourse on Pearson's views of heredity for historians of science.[9] In this chapter, I intend to create such a discourse. I will argue that whilst both Pearson and Weldon did not accept the generality of Mendelism, they both attempted, nevertheless, to provide a reconciliation of Mendelism with biometry; furthermore, Pearson continued to look for such a synthesis *even after* Weldon's death in 1906.

Since much of the scholarship on Pearson's views of heredity has been influenced by the opinions of his arch-rivals, to the extent that their accounts of Pearson have been given greater consideration than those of Pearson himself, this has led to an unbalanced account of Pearson's and Weldon's ideas of heredity. Moreover, the historiographical tendency to link Pearson's work to Galton's law of ancestral heredity, along with his work on simple correlation and simple regression, has meant that considerably less attention has been given to Weldon's role and, in particular, to his use of Pearson's chi-square goodness of fit test for Mendelian data.[10]

Bateson and Castle arrived at their conclusions that Pearson rejected Mendelism from two different perspectives. Bateson assumed firstly, that since Pearson and Weldon thought that natural selection acted upon continuous variation (as Darwin had postulated), that their models of heredity had to be based *exclusively* on continuous variation, and he also condemned Weldon's statistical analysis of Mendel's data when Weldon used Pearson's chi-square goodness of fit test. Castle formed his view by criticising the statistical results Pearson obtained when using Galton's Law of Ancestral Heredity (i.e., multiple correlation). Pearson's positivism has also been considered as a factor that would have "predisposed him to oppose Mendelism".[11]

Whilst Pearson's early ideas on heredity were influenced by Galton's Law of Ancestral Heredity (which was, indeed, underpinned by continuous variation) and in addition Pearson and the biometricians undertook two very extensive hereditarian studies of what Pearson termed "homotyposis" – these were by no means the only views of inheritance that Pearson had in his lifetime. Pearson had, in fact, begun to consider the role of discontinuous variation for problems of particulate inheritance *before* the end of the nineteenth century; moreover, his published papers and letters reveal that by the end of 1903 he began to incorporate Mendelism as a mode of inheritance for discontinuous variation, and in 1904 he had "accepted the fundamental idea of Mendel".[12] Additionally, by 1909 he suggested a synthesis of biometry and Mendelism by showing that the gametic correlations in a Mendelian population mating at random were very close to those determined for somatic correlations in a biometric investigation; thus, "there remain[ed] not the least antimony between the Mendelian theory and the Law of Ancestral Heredity'.[13]

Third wrangler in the Mathematics Tripos in 1879, Pearson's world was shaped during his formative years at Cambridge. The Mathematics Tripos, which emphasised applied mathematics as a pedagogical tool for obtaining the truth, encouraged Pearson to search for the "truth" by applying mathematical models to a variety of problems.[14] He was not thus interested in a physiological mechanism of heredity. Instead, he attempted to make sense of various hereditarian models by placing them in a mathematical context.

Francis Galton, who abandoned his medical studies at Birmingham after he inherited his father's money and managed to get a Third class in the Mathematics Tripos at Cambridge, upheld a theory of blending inheritance, and he thought that heritable variation was continuous and *normally* distributed. Though he attempted to deal with what he referred to as "non-blending inheritance" (which he sometimes called "particulate inheritance"), this discontinuous variation was analysed by using statistical methods for continuous variation (i.e., simple correlation and simple regression). Galton's theory of ancestral inheritance thus incorporated blending and non-blending inheritance. He referred to characters that did not blend (such as eye colour) as "alternative inheritance". Nevertheless, it was Galton's statistical approach to heredity that allowed him to move away from the sterile approach of using developmental and embryological ideas of heredity. His use of statistics enabled him to place problems of heredity within a population and not just within individual acts of reproduction. Pearson's views on inheritance differed from Galton's in two respects: firstly, Pearson thought most of the heritable material could be found in the parents and thus he did not attach the same amount of importance to ancestry as did Galton, and secondly, he did not think that inherited variation should be *necessarily* normally distributed

Whilst Pearson used parametric models of correlation for continuous variation in the 1890s, he had also begun to consider the role of discrete or discontinuous variation. By 1899 he had devised the phi-coefficient and the tetrachoric correlation coefficient as non-parametric statistical methods to measure relationships for discontinuous or discrete variation. In 1901 Weldon used Pearson's chi-square goodness of fit test (which Pearson had devised in 1900) for Mendel's distributions of the common garden pea, and in 1904 Pearson introduced the chi-square test of association for contingency tables (now more commonly known as the chi-square statistic), which was used to analyse Mendelian discrete data (i.e., the alleles).

Mendelian Distributions and The Chi-Square Goodness Of Fit Test

At the end of October 1900 and four months after Pearson published his paper on the chi-square goodness of fit test, Weldon wrote to Pearson that

> among pleasanter things, I have heard of and read a paper by one, Mendel, on the results of crossing peas, which I think you would like to read. Results indicate exclusive [i.e., particulate] inheritance with a very high parental *r*. It seems a good starting point for further work. It is in the *Abhandlungen des naturforschenden Vereines in Brünn* for 1865. I have the R.S. copy here, but I will send it to you if you want it.[15]

Eleven days later Weldon was able to get the full paper on Mendel's peas by Tchermak which showed clearly that there was "very great variation in the colour of pure races".[16] William Bateson had by then mentioned to Weldon that "the seeds of cross-fertilised flowers are always yellow or 'dominant' in character".[17] Weldon then wrote to Pearson that he could not see how Bateson's interpretation

> leads to Mendel's final results, which is that if self-fertilisation occur in the offspring of a cross-fertilised plant, the results is the production of *three* sets of plants, an apparently pure yellow set, an obviously hybrid set and an apparently pure green set.[18]

Throughout the summer and early autumn of 1900, Weldon was re-examining Mendel's data and considering various theories of inheritance.

At the end of October, Weldon asked Pearson for his assistance in "calculating the chances against the observed distribution" (since he wanted to use Pearson's chi-square goodness of fit test).[19] Weldon analysed Mendel's results of three of the seven set of discrete characters of the common garden pea (*Pisum sativum*) on the assumption of phenotypic dominance and independent assortment.[20] After calculating the chi-square test on Mendel's data, Weldon concluded that the "chance that a system will exhibit deviations as great or greater than these from the results indicated by Mendel's hypothesis is about 0.95".[21] When considering all of Mendel's data, Weldon remarked that they were a "wonderfully good approximations to this hypothetically probable result"[22] (i.e., Mendel's data fit a Mendelian distribution exceptionally well).

Weldon then remarked that Mendel's data were in "accord so remarkably with Mendel's summary of them the chance that the agreement between observation and hypothesis would be worse than actually obtained is about 16:1".[23] Weldon was so perplexed that the agreement between the experimental distribution of Mendel's peas against a hypothetical Mendelian distribution were so close, that he wrote to Pearson, "remembering the shaven crown, I can't help wondering if the results are too good?"[24] Though Weldon crossed this line out in the letter, he went on to write, "I do not see that the results are so good as to be suspicious".[25] So perplexing were these conclusions that a month after Weldon had begun to analyse Mendel's distribution of peas, he wrote to Pearson that Mendel "is either a black liar or a wonderful man". Though Weldon did not think the results were so good as to be suspicious, he could "see no alternative to the belief that Mendel's laws are absolutely true for his peas and absolutely false for Laxton's while those of Tchermak's are intermediate".[26]

Weldon thus wanted to know "whether the whole thing was a damned lie or not".[27] After he had read Bateson's translation of Mendel, Weldon found that he was "struggling to avoid a tendency to disbelieve the whole thing because Mendel was a Roman priest".[28] After Weldon examined all of Mendel's data and had read articles by the many who worked at first on Mendel without reworking Mendel, he was certain that Mendel "cooked his figures, but that he was *substantially* right".[29] Though Mendel's peas indicated that this was a case for particulate inheritance, Weldon was also concerned about a situation when inherited characters merge into completely blended inheritance or when the two occur together.

Weldon did not question Mendel's integrity or the results, but did challenge the interpretations and universality of the findings, though he regarded the role of ancestry as essential. He wanted to test statistically how well Mendel's results "fitted" Mendel's theoretical expectations of the 3:1 ratio (or what is also referred to as a Mendelian distribution). Weldon's findings from the chi-square goodness of fit test were never discussed explicitly by Bateson or by later generations of biologists and historians of science who viewed Weldon as an opponent of Mendel. Though R.A. Fisher alluded to the 16:1 ratio in Mendel's original results in a lecture to the Cambridge Eugenics Society in 1911 and to the 0.95 per cent when Fisher used the chi-square goodness of fit test on Mendel's data in a paper in 1936, Weldon's results never became a part of the discourse in this debate, particularly by historians of science.[30]

The statements in Weldon's papers that provoked the greatest reaction, and subsequently became the focus of much controversy, were his views on the importance of Galton's ancestral inheritance. Weldon concluded that is was not possible to regard dominance as a property of any character from a simple knowledge of its presence in one or two individual parents.

At the end of December 1901, Weldon was trying to measure the characters of gametes by using Pearson's multiple correlation. Weldon found, however, that the statistical process could be cumbersome when trying to determine the parental character that was connected to the gametes.[31] By using multiple correlation, Weldon was attempting to make sense of Mendel's data by incorporating Galton's "Law of Ancestral Inheritance" into Mendelism. Moreover, Weldon's letters to Pearson on matters of Mendel indicated that Weldon was increasingly emphasising statistical processes to interpret Mendel rather than looking for a physiological mechanism. Weldon had, of course, found great success when using Pearson's statistical methods of curve fitting to detect empirical evidence of natural selection during the previous eight years. Thus, he seemed to persist in using Pearson's methods in lieu of considering possible physiological explanations of Mendelism – perhaps hoping to replicate his earlier success.

Bateson had received Weldon's paper on Mendel on 8 February 1902 and one month later, Bateson published his fiercely polemical 100 page chapter in his book on *Mendel's Principles of Heredity. A Defence*; this chapter was written explicitly to "defend Mendel from Professor Weldon".[32] Bateson regarded Weldon's criticisms of Mendel as "baseless and for the most part irrelevant".[33] Bateson then concluded that "every case therefore which obeys the Mendelian principle is in direct contradiction to the proposition to which Professor Weldon's school is committed".[34]

Weldon thought that Bateson was being simply abusive and that his abuse went rather far beyond permissible limits. He found the whole affair to be "paltry and dirty beyond measure".[35] Weldon was even concerned that Pearson might want to remove him as one of the editors of *Biometrika*. Shortly after Bateson's book was published Bateson wrote a "most disgusting and fulsome" letter to Pearson asking him to "chuck Weldon over-board and take a certain Cambridge naturalist – Bateson – in his place!".[36] As far as Pearson was concerned if there were "ever a man [who] stood in need of horse-whipping it was the writer of that letter whom [he] had not spoken to for ten minutes in [his] life!".[37]

A couple of months later Pearson wrote to George Udny Yule that he did not see how the truth of Mendel could be tested without independent experiment and that much of Bateson's work was open to a variety of interpretations. Pearson was thus quite prepared to personally find out if there was truth in Mendel. Bateson's polemical approach to Mendelism made it difficult for Pearson to trust Bateson. Pearson argued that an "understanding of Mendel must be done by a man who does not become vulgarly abusive in a purely scientific discussion."[38]

Nevertheless, Bateson's reaction helped to cement the belief that Weldon rejected Mendelism. Thus, not only were later generations of biologists critical of Weldon's paper, but, subsequently, a number of historians of science shared their views.[39] The biometricians' focus on measuring continuous variation in populations is not surprising since so much of the variation observed within large populations *is* of a continuous nature. Mendel's discontinuous characters may have held the key to a new and far more fruitful approach to heredity, but his laws had no immediately obvious application in the many cases where a species exhibits a continuous range of variability.[40]

Weldon argued that it was possible for such discrete variables as colour and stature to be treated as continuous variables. If colour were measured on a spectrum (rather than categorised as "green" and "yellow") and height measured in inches (instead of "short" and "tall"), these variables would become continuous. Since Bateson had not addressed the role of continuous variation for problems of inheritance, Weldon found Bateson's view of inheritance to be especially problematic. Weldon's adherence to Galton's Law of Ancestral Heredity meant that he would never accept the generality of Mendelism, instead he thought that Mendelism could be used for situations where there was a clear case of discontinuous variation.

Whilst Weldon found Mendel's results perplexing in 1900, later generations of geneticists were equally perplexed. Thus, 36 years after Weldon first analysed Mendel's results, R.A. Fisher re-examined this data and he also used Pearson's chi-square goodness of fit test. Fisher's analysis was based on Mendelian gametic ratios as well as bifactorial and trifactorial experiments. Fisher found that

> a χ^2 [goodness of fit] of only 2.8110 [with eight degrees of freedom] – almost as low as the 95 per cent. Point ... was strongly significant and so low a value could scarcely occur by chance one in 2000 trials. There can be no doubt that the data from the later years of the experiment have been biased strongly in the direction of agreement with expectation.[41]

Although Fisher's results were identical to Weldon's, Fisher's conclusions elicited a very different set of responses from Weldon's. Though Fisher, no doubt, read Weldon's paper of 1902, he did not mention that he had undertaken the same statistical and experimental tests that Weldon had some 32 years earlier. Regarding Weldon's omission in Fisher's 1936 paper, Anthony Edwards commented recently that:

> naturally Fisher should have referred to Weldon (1902) in 1936. But we know that he put together the paper over the Christmas vacation in response to a request from [Douglas] McKie, a colleague at University College [London], for a contribution to his new journal *Annals of Science* ... sitting at home, he did not have Weldon's paper by him and presumably he forgot about it. [42]

Fisher thought that no explanation could be expected and that possibly "Mendel was deceived by some assistant who knew too well what was expected".[43] He presumed Mendel was aware of the independent inheritance of seven factors to have chosen seven pairs of varieties. Fisher concluded that Mendel may have thought out thoroughly the theoretical consequences of his system.[44] Fisher's results were of interest to Sewall Wright who repeated [Pearson's] χ^2 [goodness of fit] test [in 1966] from an independent tabulation and came out with substantially the same result as Fisher. There is no question that the data fit the ratios much more closely than can be expected from accidents of sampling.[45]

Twenty years after Wright published his paper, Anthony Edwards used Pearson's chi-square goodness of fit test on Mendelian segregations of all seven characters of *Pisum sativum*. His results indicated (as did those of Weldon, Fisher and Wright) that "the segregation are in general closer to Mendel's expectation than chance would dictate".[46] Whilst Edwards acknowledged that Weldon "subjected [Mendel's data] to a statistical analysis using probable errors", he does not specify that Weldon used Pearson's chi-square goodness of fit test, which so many geneticists subsequently used on Mendel's data.[47] Thus, despite the many repeated attempts to use Pearson's chi-square goodness of fit test on Mendel's data, priority has never been explicitly assigned to Weldon who was, it may now be seen, the first person to do so in his paper of 1902.

Yule's Synthesis of Mendelism and Biometry

A couple years after Weldon published his paper on Mendel in 1902, George Udny Yule offered a synthesis of Mendelism and Biometry.[48] Yule wanted to examine whether continuous variation in the phenotype could arise from changes of the genotype either due to "continuous variation of the element in the germ cell … or [from] the compounding in some way of discontinuous variation of a number of elements".[49] Yule could see how the Mendelians and the biometricians were looking at the same problem differently and thus coming up with two different approaches. The principal distinction he made was that the biometricians were interested in the phenomena of heredity within the race and thus with aggregates or groups of the population and not with single individuals. The early Mendelians, however, were interested in the phenomena of hybridisation that occurred on crossing two races that were admittedly distinct.

Yule further explained that Galton's law is only stated as an average or statistical law, and the "one quarter" contributed by the grandparents on the average might be made up by some contributing one half and others contributing nothing; the average of a series of quantities may exhibit sensible continuity of variation, even though the quantities averaged vary by discrete steps. However, as Mendelism and biometry were related "they could not be absolutely inconsistent" with each other as Bateson argued. Yule thus argued that the Law of Ancestral Heredity was a law of nature of wide generality that could not be dismissed in such a fashion. Thus Yule concluded, "Mendel's Laws and the Laws of Ancestral Heredity are not necessarily contradictory statements, but are perfectly consistent the one with the other and may quite well form part of one homogenous theory of heredity".[50]

Yet by April 1903, *Nature* thought that there must be something to Pearson's approach to heredity. On 9 April various articles from *Biometrika* were reviewed in *Nature*. The recent work on Mendel was of particular interest, and the reviewer wrote that the last three numbers continue to record results of high biological interest. The excellence of Prof. Karl Pearson's elaborate studies in statistical theory is becoming widely recognised, and his comments and criticism add much to the value of the work of other contributors. [51]

It was concluded that even "Mr Bateson, at all events, is not disposed to admit that the facts so far obtained are discordant with Mendel's law, but it must be allowed that much of the evidence is *prima facie* in favour of ancestral inheritance".[52]

In the summer of 1904, The British Association for the Advancement of Science held their annual meeting at Cambridge. On 18 August, Bateson delivered his presidential address for the Zoology Section which gave him the opportunity to attack the work of Pearson and Weldon by arguing that the

gross statistical method is a misleading statement; and applied to these intricate discrimination, the imposing Correlation table into which the biometrical Procrustes fits his arrays of unanalysed data is still no substitute for the common sense of a trained judgement.[53]

Bateson went on to say that "in direct contradiction to the methods of current statistics, Mendel [would have said] that masses must be avoided".[54] As far as Bateson was concerned "breeding gave the only test".

On the following day, Weldon discussed his results on the colour of cotyledon in peas, and his student, Arthur Darbishire, gave an account of some of his experiments on the breeding of mice. Bateson replied in some detail to Weldon's criticisms and "maintained that by the Mendelian hypothesis alone was it possible to draw together the vast number of observed facts which had seemed utterly incoherent".[55] Pearson then replied that "the introduction of [biometrical] methods of precision had nothing to do with Mendelism or ancestral law".[56] Pearson found it troubling that "the Mendelians produced figures without making any attempt to show that the figures were consonant with the theory". He suggested a truce since "controversy could only be settled by investigation, not disputation".[57] Professor Hubrecht hoped, however, "that the controversy would continue ... [since] interest in this important inquiry was greatly quickened by the controversy".[58] The Rev. T.R.R. Stebbing then remarked that "you have all heard ... what Professor Pearson has suggested ... but what I say is let them fight it out".[59] Whilst Pearson's conciliatory mood had little effect on this debate, the "excitement of the meeting seemed to have braced Weldon to greater intellectual activity".[60]

A couple of months after the meeting, Pearson wrote a note to *Nature* to indicate that he accepted Mendelism and to clarify the role of biometry in Mendel he stressed that

biometry is only the application of exact statistical method to the problem of biology. It is no more pledged to one hypothesis of heredity than to another, but it must be hostile to all treatment which uses statistics without observing the laws of statistical science ... for I thought and still think Mendel himself considered "dominance" as an important part of his system.[61]

Pearson argued further that Weldon and he had made

> the only attempt to carry out any form of Mendelianism to its logical
> conclusions ... . Notwithstanding that in every generation dealt with in
> my memoir [on Mendel in 1904] *the fundamental idea of Mendel is accepted*
> and the re-crossing of the parental forms with each member of the
> generation occurs and is treated as giving its Mendelian result.[62] (Italics
> mine.)

If the "assumption made is that a Mendelian character is a discrete unit" then
Pearson maintained that the biometricians were "absolutely the first to apply
[statistical] methods [for discrete variables] treating the Mendelian theoretical
characters as units".[63] One of Weldon's earliest attempts to synthesise
Mendelism with biometrical methods can be found in his Oxford lecture notes
on heredity. He began by looking at a set simple of correlations over several
generations by examining the correlation between two generations at a time, and
he then calculated Pearson's multiple correlation to determine the contribution
of each generation on the inherited character. Pearson also showed in 1904, in
his Proposition II, the stability of the 1AA:2Aa:1aa ratio in a population mating
at random.[64] This equilibrium principle was dealt with more fully by Hardy and
Weinberg in 1908.[65]

To a large extent, the highly polemical nature of this debate ended on 13
April 1906 when Weldon died of double pneumonia.[66] The news of Weldon's
death, at the age of 46, was a shock to the scientific community and no one
more than Pearson felt the loss so poignantly: he had not only lost "the closest
friend he ever had", but he found himself, at once, alone in the scientific
community with no one to guide him with such biological problems as
Mendelian inheritance and no one to share his enthusiasm for his statistical
work.[67] The loss was also felt deeply by Weldon's one time colleague, and later
arch-rival, William Bateson, who wrote to his wife Beatrice,

> To Weldon I owe the chief awaking of my life. It was through him that I
> first learnt that there was work in the world which I could do. Failure
> and uselessness had been my accepted destiny before. Such a debt is
> perhaps the greatest that one man can feel towards another; nor have I
> been backward in owning it. But this is the personal, private obligation
> of my soul.[68]

Neither Pearson nor Weldon were prepared to accept Bateson's view of
Mendelian inheritance, which not only rejected biometry, but also rejected the
inheritance of those characteristics which would be classified as "continuous".
In 1912, Pearson wrote a paper that represented his concerted attempt to
synthesise Mendelism with biometry, but the Mendelians did not respond to
Pearson. Some six years later, in 1918, R.A. Fisher showed that Mendelism was
compatible with continuous variation by demonstrating that "the statistical
properties of any feature [could be] determined by a large number of Mendelian
factors".[69]

Since Pearson had no contemporary from whom to seek aid or advice, it
seems that he alone was not prepared to modify Weldon's views to incorporate
Fisher's work. Pearson's reluctance to modify his views may well have had an

emotional basis. That is, if Pearson were to incorporate a newer view, he would have had to relinquish Weldon's views and this may have very likely led to a near abandonment of the spirit of Weldon which Pearson tried to kept alive, in part, by publishing Weldon's unfinished work on Mendelian inheritance until about 1932 (four years before his own death).

To conclude, by 1903 both Pearson and Weldon incorporated Mendelism as a theory of inheritance for discontinuous variation, and in the following year neither of them saw any antagonism between Mendel and Galton's Law of Ancestral Inheritance (which signified to Pearson the statistical underpinnings of multiple regression), and the two were shown to be compatible. Bateson was thus wrong to say that neither Weldon nor Pearson accepted Mendelism, but he was right to argue that, for Weldon, Mendelism was irrelevant for most cases.

Pearson's efforts to synthesise Mendelism with biometry, three years after Weldon's death, were not acknowledged by the Mendelians. Hence, the claims made by Punnett, Hogben, Huxley, Sturtevant, and Wright as well as those historians of science who believed that Pearson opposed Mendelism, and that he alone delayed scientific progress for 20 years, have not fully addressed the totality of the hereditarian views of Pearson and Weldon. Moreover, the historiographical tendency to oversimplify the struggle between Mendelism and biometry by attributing blame to Pearson has not addressed a variety of factors and complex motives in the scientific community at that time. After all, it was only in the 1920s when it became possible "to create a science of population genetics by using the biometricians' statistical techniques to apply the laws of genetics to the more complex cases of heredity in whole populations".[70]

The triumphalist and positivist image of scientists, used by the early Mendelians in particular, inhibited the possibility of understanding the biometricians' analysis of Mendelism. The claims made by the early Mendelians also demonstrate how a community of scientists, who either ignored some of Pearson's papers or misrepresented Pearson's and Weldon's views on Mendelism – perhaps as a result of not understanding or misinterpreting the Victorian mathematical-statistics of Pearson – were able to create the belief that Pearson and Weldon opposed or rejected Mendelism (and make further claims that Pearson "hindered scientific progress").[71] An examination of the totality and the complexity of the hereditarian-statistical work of Pearson and Weldon has, however, made it possible to redress the balance in the historiography of their views on inheritance and Mendelism in particular. This chapter has thus shown that whilst neither Pearson nor Weldon accepted the generality of Mendelism, they did not reject it completely; moreover, Pearson made a serious attempt to reconcile the theory with his own techniques.

References

1   Peter Bowler, "The Rediscovery of Mendelism" in Robert A. Peel and John Timson (eds.), *A Century of Mendelism* (Galton Institute: London, 2000), pp. 1-14.
2   *Ibid*, p1 .
3   See Peter Bowler, *The Mendelian Revolution. The Emergence of Hereditarian Concepts in Modern Science and Society* (London, 1984), pp. 103, 112; Augustine Brannigan, The

Reification of Mendel. *Social Studies of Science*, 9 (1979), 422-9; Robert Olby, Mendel No Mendelian. *History of Science* 17 (1979), 53-72 reprinted in *Origins of Mendelism* (London, 2nd ed. 1985), p. 253.

4 *Ibid.*

5 Eileen Magnello, Karl Pearson's Gresham Lectures: W.F.R Weldon, Speciation and the Origins of Pearsonian Statistics. *British Journal for the History of Science* 29 (1996). 43-63. For a fuller account of Pearson's establishment of various departments and laboratories at UCL see, *Idem*, The non-correlation of biometrics and eugenics: Rival forms of laboratory work in Karl Pearson's career at University College London. *History of Science*, 37 (1999), 79-106, 123-50.

6 William Bateson, *Mendel's Principles of Heredity: A Defence* (Cambridge University Press, 1902), pp. 104-208. William Ernest Castle, The laws of heredity of Galton and Mendel, and some laws governing race improvement by selection. *Proceedings of the American Academy of Arts and Science*, 39, No. 8 (November 1903), 224-232.

7 J.B.S. Haldane, Karl Pearson in *Speeches Delivered at a Dinner Held at University College London on the Occasion of Karl Pearson's Centenary Celebration*, 13 May 1957 (privately issued by the *Biometrika* Trustees, 1958); Lancelot Hogben, *Statistical Theory: The Relationship of Probability, Credibility and Error* (London, Allen and Unwin, 1957), p. 235; Julian Huxley, *Evolution: The Modern Synthesis*, 2nd ed., (London, Allen and Unwin, 1963), p. 24; Reginald Punnett, Early days of genetics. *Heredity*, 4 (1950) 2-10; A. H. Sturtevant, *A History of Genetics* (New York, Harper and Row, 1959), p. 58; and Sewall Wright, "The foundations of population genetics or Evolution in Mendelian populations". *Genetics* 16 (1934), 98.

8 Joan Fisher Box, R. A. Fisher. *The Life of a Scientist.* (John Wiley: New York, 1978), p. 3.; Bernard Norton, Biology and Philosophy: The Methodological Foundations of Biometry. *Journal of the History of Biology* 8 (1975), 85-93; Olby, (endnote 3), p. 129; William Provine, *The Origins of Theoretical Population Genetics* (Chicago, University of Chicago Press, 1971), p. 69; R.G. Swinburne, Galton's Law: Formulation and Development. *Annals of Science* 24 (1965), 227-46.

9 P. Froggatt and N.C. Nevin, The 'Law of Ancestral Heredity' and the Mendelian-Ancestrian Controversy in England 1889-1906. *Journal of Medical Genetics* 8 (1971), see especially pp. 20-4.

10 See, for example, Swinburne (endnote 8); Provine (endnote 8); Norton (endnote 8); Donald Mackenzie, *Statistics in Britain 1865-1930: The Social Construction of Scientific Knowledge* (Edinburgh, Edinburgh University Press, 1981).

11 J.B.S. Haldane (endnote 7), p. 9; Norton (endnote 8), pp. 87-88; Olby (endnote 3), p. 129.

12 Karl Pearson, Mathematical Contributions to the Theory of Evolution. XII. On the generalised theory of alternative inheritance with special references to Mendel's laws. *Philosophical Transactions of the Royal Society* A 203 (1904), 53-86. Pearson read the paper on 26 November 1903. *Idem*, Mendel's Law. *Nature*, 70 (27 October 1904). 626-7.

13 *Idem*, The theory of ancestral contributions of a Mendelian population mating at random. *Proceedings of the Royal Society* 81 (1909), 225-9.

14 For a fuller account of the nineteenth century mathematics training see Andrew Warwick, Exercising the student body. Mathematics and Athleticism in Victorian Cambridge in Chris Lawrence and Steven Shapin (eds.), *Science Incarnate: Historical Embodiments of Natural Knowledge* (Chicago, University of Chicago Press, 1998).

15 W.F.R. Weldon, Letter to Karl Pearson, 16 October 1900, KP:UCL/891/1.

16 W.F.R. Weldon, Letter to Karl Pearson, 27 October 1900, KP:UCL/891/1.

17 W.F.R. Pearson, Letter to Karl Pearson, 5 December 1900, KP:UCL/891/1

18 *Ibid.*

19 W.F.R. Pearson, Letter to Karl Pearson, 27 October 1900, KP:UCL/891/1

20 W.F.R. Weldon, Mendel's Law of Alternative Inheritance in Peas. *Biometrika*, 1 (1902), 228-54. Froggatt and Nevin (endnote 9) thought that Weldon was 'on good ground as the dominance by which Mendel selected his seven differentiating characters was even then under challenge and Mendel's selection was considered to be a combination of coincidence and superficial examination', p. 18.

21 Weldon (endnote 20), p. 235.

22 W.F.R. Weldon, Letter to Karl Pearson, 29 October 1901, KP:UCL/891/1

23 W.F.R. Weldon, Mendel's Law of Alternative Inheritance in Peas. *Biometrika*, 1 (1902), 233.

24 W.F.R. Weldon, Letter to Karl Pearson (n.d. November 1901), KP:UCL/891/1.

25 *Ibid.*

26 *Ibid.*

27 *Ibid.*

28 W.F.R. Weldon, Letter to Karl Pearson, 13 November 1901, KP:UCL/891/1.

29 W.F.R. Weldon, Letter to Karl Pearson, 28 November 1901, KP:UCL/891/1.

30 Bernard Norton and Egon Pearson, "A note on the background to, and referring of, R.A. Fisher's paper [On] the correlation of relatives on the supposition of Mendelian experiments". *Notes and Records of the Royal Society* **31** (1976) p.152. Also cited in A.W.F. Edwards, Are Mendel's Results Really too Close? *Biological Review* 61 (1986), p. 295. Though Anthony Edwards commented in this paper 'Fisher's remark about the odds of 16 to 1 was probably taken directly from Weldon's [paper of] 1902,"

31 W.F.R. Weldon, Letter to Karl Pearson, 18 December 1901, KP:UCL/891/1.

32 William Bateson, *Mendel's Principles of Heredity: A Defence* (Cambridge: 1902) p. viii.

33 *Ibid*, p. 106.

34 *Ibid*, p. 114.

35 W.F.R. Weldon, Letter to Karl Pearson, 3 June 1902, KP:UCL/891/2.

36 Karl Pearson, Letter to G. Udny Yule, 27 August 1902, KP:UCL/905.

37 *Ibid.*

38 Karl Pearson, Letter to G. Udny Yule, 27 August 1902, KP:UCL/905.

39 Punnett (endnote 7) wrote of 'Weldon's belittlement of Mendel's work in the first volume of *Biometrika*', p. 3; Provine (endnote 8) remarked that 'Weldon initiated the attack upon Mendelian inheritance in the second number of *Biometrika*', p. 70. Brannigan (endnote 3) thought that 'the initial reaction toward Mendel's paper among the biometricians was negative', p. 434.

40 Bowler (endnote 3), p. 71.

41 R.A. Fisher, Has Mendel's Work Been Rediscovered. *Annals of Science* 1 (1936) p. 130.

42 Anthony Edwards (personal correspondence, 5 November 2001).

43 *Ibid*, p. 132.

44 Thus Peter Bowler remarked that 'since the pea has seven chromosomes it was supposed that Mendel chose the maximum number of independent segregating characters available to him'. Bowler (endnote 3), p. 102.

45 Sewall Wright, Mendel's Ratios in Curt Stern and Eva R. Sherwood (eds.), *The Origins of Genetics. A Mendel Source Book* (San Francisco, W.H. Freeman, 1966), p. 173.

46 A.W.F. Edwards, Are Mendel's Results Really too Close? *Biological Review* 61 (1986), p. 295-312.

47 *Ibid.,* pp. 295-6.

48 George Udny Yule, Mendel's laws and their probable relations to intra-racial heredity. *New Phytologist* 1 (1902), 193-207, 221-237.

49 *Ibid,* p. 235. The *New Phytologist* was a new botanical journal set up to 'establish easy communication' and was not intended to be the rival of any existing periodical. Also see Froggatt and Nevin (endnote 8), p 18.

50 *Ibid.*

51 Anon, Scientific Serial. *Biometrika, Nature* 67 (9 April 1903), 550.

52 *Ibid.*

53 William Bateson, Presidential Address, in Beatrice Bateson, *William Bateson* (Cambridge, Cambridge University Press, 1928) p. 240.

54 William Bateson, " Opening address [Section D, Zoology] by William Bateson, M.A., F.R.S., President of the Section, *Nature* 70, (August 25, 1904), 409.

55 W.J.S.L., The British Association and referees. *Nature* 70 (29 September 1904), 538-9.

56 *Ibid,* p. 539.

57 *Ibid.*

58 *Ibid.*

59 Punnett (endnote 7), pp. 7-8.

60 Karl Pearson, Mendel's Law, *Nature* 70 (27 October 1904), 626- 7.

61 *Ibid,* p. 626-7.

62 *Ibid,* p. 627.

63 *Ibid,* p. 210.

64 *Ibid,* p. 626-7.

65 G.H. Hardy, Mendelian proportions in a mixed population. *Science* 28 (1908), 49-50. C. Weinberg read his results at the Society for Natural History in Stuttgart some six weeks before Hardy's paper was published. See Curt Stern, The Hardy-Weinberg law. *Science* 97 (1943), 137-8.

66 Florence Weldon, Telegram to Karl Pearson, 13 April 1906, KP:UCL. Pearson wrote the time and cause of death on the telegram.

67 Karl Pearson, Letter to Mrs. Weldon, 29 April 1906, KP:UCL.

68 William Bateson, Letter to Beatrice Bateson, in Beatrice Bateson, *William Bateson, F.R.S. Naturalist. His Essays and Addresses Together with a Short Account of His Life.* (Cambridge, Cambridge University Press, 1928), p. 103.

69 R.A. Fisher, The correlation between relatives on the supposition of Mendelian Inheritance. *Transaction of the Royal Society of Edinburgh* 52 (1918), 432. According to William Provine, Castle had suggested that continuous variation of heritable traits might have a Mendelian interpretation. In William Provine, *Sewall Wright and Evolutionary Biology* (Chicago, University of Chicago Press, 1986) p. 37.

70 Bowler (endnote 3), p. 120.

71 Jerry Ravetz has referred to this process of ignoring or misinterpreting scientists' papers as the 'social construction of ignorance'. See Jerome Ravetz, The Sin of Science. *Knowledge: Creation, Diffusion, Utilization* 15 (2 December 1993), 157-65. *Idem,* A leap into the unknown. *The Times Higher Education Supplement* (26 May 1993), 18-9.

4. Mendelism and Man 1918–1939

A.W.F. Edwards

Introduction

1918 and 1939, the year of peace and the year of war, are best remembered in human genetics as, respectively, the year in which the Hirszfelds demonstrated that the frequencies of the ABO blood groups varied from one population to another and the year in which P. Levine and R.E. Stetson uncovered the mechanism leading to haemolytic disease of the newborn, soon shown to be a consequence of maternal-foetal incompatibility at the newly discovered Rhesus blood-group locus. The first observation heralded the birth of anthropological genetics which, through the efforts of W.C. Boyd and A.E. Mourant, and in our own day L.L. Cavalli-Sforza, has made such a contribution to our knowledge of the recent evolution of man. The second observation was important for clinical genetics, for although transfusion reactions had been understood for some time, here the Mendelian basis of a clinical problem was laid bare, leading eventually to its solution.

Between the two world wars much of human genetics as it existed before the arrival of cytogenetics in the 1950s was created, mostly, as we shall see, in connection with the blood groups. The over-eager application of a naive Mendelism which had characterised the first two decades of the century was replaced by the realisation that much had to be learnt before any benefits could be expected from the new subject.

Whilst assembling material for my talk, I took from my shelves the Proceedings of the Symposium held to celebrate the fiftieth anniversary of William Bateson's foundation of the Genetical Society in 1919. It was a poignant read of the early history of the Society, the more so since I have just been involved in trying (and failing) to persuade the Society not to oust Bateson's adjective "Genetical" from its name and replace it with his noun "Genetics". My membership has overlapped with at least one founder-member, the indomitable J.B.S. Haldane, whom I first saw at one of its meetings when I was a research student.

At the Society's anniversary symposium, our chairman this afternoon, Professor J.H. Edwards, gave a paper, reproduced in full in the Proceedings, which was (if he will allow a fraternal compliment) a masterly survey. Entitled "The Application of Genetics to Man – 1869-1969" (1869 had seen the publication of Francis Galton's *Hereditary Genius*), it is an important document for historians who wish to ask "What did people think in 1969 had been the achievements between the wars?"

Professor Edwards starts with R.A. Fisher's 1918 paper, which I will come back to in a moment, succinctly describing it as having "bridged the gap between observations at one locus and inferences about many". He continues, "In the twenties, when genetics had a society as well as a name, human genetics was curiously inactive, partly, it would seem, through the propaganda of the

Eugenics Society which repelled men, such as Bateson, by an evangelism conflicting with their standards of truth". I think this statement needs qualifying in that the Eugenics Society in the 1920s and 1930s was the nearest thing there was to a Human Genetics Society, and that although it might have been too much for Bateson it became a catalyst for research and a financial supporter of the struggling young subject.

We will be hearing more about William Bateson later, but his attitude – and his vigorous mode of expression – are well exemplified by a comment in his 1912 Herbert Spencer Lecture: "... but if we picture to ourselves the kind of persons who would infallibly be chosen as examples of 'civic worth' – the term lately used to denote their virtues – the prospect is not very attractive. We need not for the moment fear any scarcity of that class, and I think we may be content to postpone schemes for their multiplication".

"In the thirties", continued Professor Edwards, "new blood groups were discovered and found to 'mendelize' consistently and their alleles to show marked variations in proportion by race; the mutation rate to dominant and X-linked disorders in man was determined by both direct and indirect methods by [L.S.] Penrose and Haldane: subsequent studies on other species, and other conditions in man, have been very consistent with these early estimates. [Lancelot] Hogben's *Nature and Nurture*, perhaps the most influential contribution of the thirties to Human Genetics, was published in 1933". I rather doubt this view of the influence of *Nature and Nurture*, but it is an extremely valuable book for the historian. It might have had greater influence had Lancelot Hogben called it simply *Human Genetics* instead of being unable to resist the Galtonian – indeed Shakespearean – phrase "nature and nurture", but it was after all intended partly as a critique of Fisher's 1918 paper, which had purported to provide a calculus for estimating the two components.

Advances besides Linkage Estimation

Much of my account will be concerned with the origin of methods for detecting and estimating linkage in man, but I shall first cover briefly some of the more important other developments. First, "Fisher's 1918 paper", as "The correlation between relatives on the supposition of Mendelian inheritance", is universally known. Although Fisher wrote this, the foundation work for biometrical genetics, in a specifically human context, the paper has had its greatest impact in quantitative genetics as applied to plant and animal breeding. Yet we should not forget that its *raison d'être* was the correlation between human relatives. It was in this connection that Fisher coined the word "variance" and first put forward the analysis of variance, which was to become such an important part of statistics. The paper was initially submitted to the Royal Society for publication, but was withdrawn after unfavourable reports on it by the referees, Karl Pearson and R.C. Punnett. On telling the story himself Fisher used to add, after giving their names, "both of whom I later succeeded". There is a detailed commentary on the 1918 paper by P.A.P. Moran and C.A.B. Smith, but unfortunately without much of an introduction, though it does start "Sir Ronald Fisher's 1918 paper on the correlation between relatives is one of the classical papers of scientific literature".

When Pearson thanked Fisher for an offprint of the paper as finally published by the Royal Society of Edinburgh, he wrote "Many thanks for your memoir which I hope to find time for. I am afraid I am not a believer in cumulative Mendelian factors as being the solution to the heredity puzzle". Bateson was not too keen either, writing to Fisher in 1920 "I am suspicious of the value of quantitative 'traits' at this stage of genetics". Hogben devoted a chapter of *Nature and Nurture* to Fisher's paper and its implications and was the first to point out that its actual deductions from data were dubious because Fisher's brilliant analysis, though it encompassed dominance and epistacy, neglected environmental within-family correlations.

Secondly, mention must be made of Fisher's *The Genetical Theory of Natural Selection* published in 1930. Apart from being one of the most important scientific books of the twentieth century ("arguably the deepest and most influential book on evolution since Darwin" – according to Jim Crow), *The Genetical Theory* is explicitly human in its context, and not only in the last five chapters on man. It has already acquired a substantial secondary literature, and my recent article on it in J.F. Crow and W.F. Dove's *Perspectives* series in the journal *Genetics* (Edwards, 2000) should be consulted to compensate for the impossibility of doing the book justice in the couple of sentences which are all there is space for here.

Thirdly, there are the great advances in mathematical genetics in the period 1922-1932, covering of course not just man, but diploid sexually reproducing organisms generally. I have recently described these in advances in another article, "Darwin and Mendel united: the contributions of Fisher, Haldane and Wright up to 1932", this time in an article in the *Encyclopedia of Genetics* edited by E.C.R. Reeve (Edwards, 2001). Haldane's long series of mathematical papers is discussed there.

One idea from a little later is that the inbreeding coefficient of an individual is the probability that his maternal and paternal genes at a locus are identical by descent and not just in type. This important concept, especially in human genetics, is usually attributed to others, but I find it in Haldane and Pearl Moshinsky's 1939 paper on inbreeding and cousin marriage. Oddly, they attribute to Sewall Wright's justly famous 1922 paper the connection between his inbreeding coefficient and this probability, but I cannot see it there. Even earlier, however, Raymond Pearl (1914) had seen what was required. He introduced a coefficient of inbreeding K based on the number of ancestors of an individual compared with the number there would have been if all were unrelated, realising that this was only a rough index of what he really wanted:

> The values of the K's for a particular pedigree evidently furnish a rough index of the probability that the two germ-plasms which unite to form an individual are alike in their constitution. This will follow because of the fact that the probability of likeness of germinal constitution in two individuals must tend to increase as the number of ancestors common to the two increases. Just what is the law of this increase in probability is a problem in Mendelian mathematics which has not yet been worked out.

Finally, to consider the American and other contributions between 1918 and 1939. One of the richest modern sources for the history of genetics is the series of *Perspectives* articles already mentioned, particularly the recent volume reprinting a large number of them (Crow and Dove, 2000), but we find very little human genetics. Fisher, Haldane and Wright naturally make their entrances, as do Penrose and Hogben and Felix Bernstein, but the treatment is mostly biographical. Wright was, of course, American, and Bernstein came to America from Germany in 1928. I am similarly unable to point to any extensive account of the German and Scandinavian contributions. Gunnar Dahlberg, an influential figure in Sweden and brave critic of wartime German practices, is not even in the index of the *Perspectives* volume. In the amazingly full "Essay on Sources" in his book *In the Name of Eugenics: Genetics and the Uses of Human Heredity,* Daniel Kevles remarked in 1985, "There is no comprehensive historical study of human genetics, and nothing more than a few autobiographical reminiscences by its practitioners". Some rectification of this, at least from a medical perspective, is afforded by Victor McKusick's 1996 article "History of Medical Genetics".

Linkage

I now turn to the estimation of genetic linkage. In his 1969 survey Professor Edwards did not mention this as an activity of the 1920s and 1930s. He did so of course in respect of the 1950s and 1960s, stating "Linkage studies in man are well advanced" and ending with the prophesy "Perhaps their biggest contribution will be in introducing the concept of likelihoods to biologists and physicians". Nearly correct, but in the event the linkage practitioners have pursued a wobbly track between Bayesian and repeated-sampling methods, and a real understanding of likelihood has grown more in the fields of phylogenetic estimation and genealogical computation. My own book *Likelihood,* originally 1972 but in print ever since, was inspired by the phylogenetic tree problem, not linkage.

It is well known that in 1919 Haldane published two papers on linkage estimation. I described them in my recent history of early linkage theory as being "often wrongly supposed to be the foundation of linkage estimation theory" (Edwards, 1997). My account runs from 1911 to 1934 and concludes, "By 1928 the statistical (but not the computational) problems of linkage estimation in experimental organisms had been solved, leaving Haldane and Fisher [in 1934] to turn to the peculiar problems of linkage estimation in man, where a start had already been made by other workers in Germany and England". Thus my history stops just where the human genetics interest starts, and I hope to continue it one day, paying special attention to Fisher's *u*-statistics.

London, 1930

Rather than delve into the details of linkage estimation theory, I shall sketch the academic background which enabled the remarkable development to take place, almost entirely in London in the 1930s.

I was fortunate enough to hear Professor Lionel Penrose's Presidential Address to the Third International Congress of Human Genetics in Chicago in 1966 and it remains a valuable portrait of what he called "the English school" of human genetics based on the Galton Laboratory, which he himself headed after the second world war. He mentions, incidentally, that Galton and A.E. Garrod must have met in 1891, for Garrod's fingerprints are in (as opposed to on!) Galton's notebook.

In October 1930 Penrose, having just taken his Cambridge M.D., was appointed to a new post at the Royal Eastern Counties' Institution in Colchester to undertake research into the causes of mental deficiency. The post had been established on the initiative of the Darwin Trust and supported jointly by them and the Medical Research Council (MRC). The Trust had at its disposal the income from a property which had been owned by Sir Horace Darwin, Charles's fifth (and last) surviving son, who had died in 1928. Horace's daughter Ruth was the instigator.

At the beginning of 1930 we find Karl Pearson still Galton Professor of Eugenics at University College and head of what was still called the Galton Laboratory of National Eugenics, R.A. Fisher still Chief Statistician at Rothamsted Experimental Station and J.B.S. Haldane Reader in Biochemistry at Cambridge, whilst Lancelot Hogben, forever on the move, had just become the first and last Professor of Social Biology at the London School of Economics (LSE), supported by the Rockefeller Foundation. "I surmise", wrote Hogben, "that [Harold] Laski's main concern in inveigling me into taking the chair of Social Biology was that the brass hats of the Eugenics Society were already congratulating themselves on the prospect of one of their co-religionists getting the job". Fisher had, in fact, applied after having corresponded with Leonard Darwin, Horace's elder brother by a year, about the possibility. Darwin thought Fisher should apply, but "you must not mind failure. They [at the LSE] are, I think, a cranky body, and one cannot guess what line they will take ...".

These, then, were the *dramatis personae*. Already we see the influence of the Rockefeller Foundation, the Medical Research Council and the Darwin family (not forgetting the Galton connection there either). Fisher's biographer, Joan Box, writes, "As early as 1924, when the Rockefeller Institute of Health was established in London, Fisher had prepared a notice for the *Eugenics Review*: 'to bring to the attention of the Ministry of Health the urgent desirability of establishing a Chair of Human Heredity in relation to disease' ...". (At about the same time, Fisher was trying to persuade Cambridge to establish a professorship of mathematical statistics; either would have suited him.)

One further event of 1930 should not go unnoticed: the Twitchin bequest to the Eugenics Society, which greatly facilitated its evolution from a propaganda society of doubtful scientific virtue into one which appreciated the need for research in human genetics and had the means to support it – the so-called "reform eugenics".

C.C. Hurst and the British Council for Research in Human Genetics

In 1931 Major C.C. Hurst wrote to a number of the most influential biological and medical scientists in Britain (including Garrod, incidentally, though he happened to be abroad) inviting them to a meeting at the London School of Economics (LSE) on 21st July to discuss the need for an initiative to promote research in human genetics in Britain. Hurst is rather a forgotten figure, but he should not be. I stumbled on his papers in Cambridge University Library when whiling away an odd moment looking in the manuscripts index to see if there were any letters from Fisher. There were two to Hurst, and when I asked for them I was brought nine boxes of Hurst correspondence, notes and typescripts. There are letters from Galton (1), Bateson (234), R.C. Punnett, Karl Pearson, Leonard Darwin, T.H. Morgan and a host of other people almost as famous. The collection is described by Rona Hurst (the second Mrs Hurst) in *The Mendel Newsletter* No. 11 for June 1975 and has been carefully annotated by her throughout.

Born in 1870, Hurst was able to follow his genetical interests through his large nursery in Leicestershire, which he inherited from his father. Prevented from going up to Cambridge by an attack of tuberculosis, he busied himself with the genetics of orchids, meeting Bateson in 1898 and again in 1899 at the International Conference on Hybridisation, reckoned now to have been the first Genetics Conference. After the rediscovery of Mendelism, Hurst turned to questions of coat colour in horses and eye colour in man, crossing swords with Pearson in the process. No one seems to have challenged Hurst's conclusion, in 1907, that blue eye colour is a Mendelian recessive, and Dr. Eiberg, of Copenhagen, tells me that the main locus is on chromosome 15. It was this example which Punnett used at the Royal Society of Medicine in 1908, eliciting the question from a member of the audience "if brown is dominant to blue, why is the population not becoming increasingly brown-eyed?" Punnett could see intuitively that there must be some kind of equilibrium, but to clear up the matter he asked his Cambridge friend, the mathematician G.H. Hardy, who replied with the Hardy equilibrium formula (now the Hardy-Weinberg formula). Hurst, like Leonard Darwin, Haldane, Punnett and, of course, Bateson, was a founder member of the Genetical Society in 1919.

The first world war and its economic aftermath destroyed Hurst's nursery business and in 1922 he moved to Cambridge and became a Research Student at Trinity, taking his Ph.D. in 1924 and an Sc.D. in 1933. He published *The Mechanism of Creative Evolution* in 1932, an excellently produced account of genetics at the time, and long a member of the Eugenics Society, he joined its Council at the time when Fisher was one of the Honorary Secretaries.

The meeting which Hurst called at the London School of Economics on 21 July 1931 was well-attended. It was chaired by Sir Daniel Hall, Director of the John Innes Horticultural Institution, and amongst those present were Sir Walter Fletcher, Secretary of the MRC and Sir William Beveridge, Director of the LSE. Hogben, Haldane, F.A.E. Crew and R. Ruggles Gates were there, but not Fisher, who was in the United States, or Pearson, who was unable to be present. On receiving Hurst's invitation Fletcher had replied "What is really wanted is the assemblage in a small committee of men with first-hand knowledge of the subject to do the scientific 'staff work' in this field. Before I heard from you at all my Council had had this matter under consideration, and were contemplating the appointment of a small committee to advise us upon our policy with regard to the better study of human inheritance, in which we have long been hoping to make a forward movement". At the meeting itself he anticipated the likelihood of MRC support, but thought attaching individual scientists to existing laboratories was a better plan than contemplating a separate research institute.

The meeting appointed a drafting committee, to include those participants named above with the exception of Sir Daniel Hall, and the addition of one or two others, with Hurst as Secretary. "What about Fisher?" an anonymous voice called out, to which Hogben had a ready answer: he was in America. This committee met on 22 September and had before them a draft "Scientific Memorandum on The Needs of Research in Human Genetics in Great Britain – An Appeal to the Rockefeller Foundation of New York". Styling themselves the "British Council for Research in Human Genetics" they approved the draft for transmission, save for a couple of paragraphs at the end. It must have been written by Hogben, for it consists principally of the whole of §4 of the last chapter of his *Genetic Principles in Medicine and Social Science*; indeed, this section reads as though it was added to the book at the last moment. It contains the statement "On the basis of such work as Bernstein's analysis of the blood groups, it is now legitimate to entertain the possibility that the human chromosomes can be mapped". Surely here is the birth of the Human Genome Project.

The MRC Committee on Human Genetics

The main effect of Hurst's initiative was to persuade the MRC that it was time to stop contemplating a committee on human genetics and act. Hogben, from his base at the LSE, seems to have been the main supporter of the idea, and the MRC's Committee on Human Genetics met for the first time on 2 March 1932. Haldane was chairman and the other members were Julia Bell (at the Galton, supported by the MRC), E.A. Cockayne (Physician to the Middlesex Hospital and a founder member of the Genetical Society), Fisher, Hogben, Penrose and J.A. Fraser Roberts (a physician who, like Penrose, was working with mental patients).

The main benefit of the Committee was to bring together this remarkable group of people for discussion. Hogben, already in London, was one step ahead of Haldane and Fisher in the quest for linkages. He had been making great strides in introducing methods for linkage analysis based on Bernstein's work, which he came across whilst writing *Genetic Principles in Medicine and Social Science* in 1931, where he devoted a chapter to it. Two Royal Society *Proceedings* papers were published in 1934. Haldane soon followed Hogben's lead, and then Fisher followed with his efficient maximum-likelihood method and Penrose with his 1935 sib-pair method. Hogben's interest then seems to have waned. He was distracted by illness (during which he wrote *Mathematics for the Million*) and in 1937 left London for Aberdeen. He published nothing on linkage after 1935.

Haldane's interest turned to the X-chromosome. It is sometimes said that in 1937 Bell and Haldane found the first linkage in man, between the two X-linked loci haemophilia and colour blindness (I omit a discussion of the precise variants now known to be involved), but this is not quite true. What they did, and did magnificently, was to estimate the recombination fraction between the two loci, with Haldane (presumably) introducing Bayesian methodology into linkage analysis, thereby using the whole of the likelihood function rather than just its maximum. By a roundabout route this has become the basis of modern methods.

However, the linkage itself had already been noticed. In 1933 Cockayne published a book, *Inherited Abnormalities of the Skin and Its Appendages*, in which he wrote, in a section on linkage: "The peculiarities of sex-linked inheritance have enabled us already to locate the genes for a number of abnormal characters in the X-chromosome, but there appears to be only one known example of linkage between two of them. Davenport [1930] has published a short pedigree which demonstrates linkage between haemophilia and red-green blindness". Penrose also knew about it, but was less sanguine, writing in his 1933 Buxton Browne Prize Essay *The Influence of Heredity on Disease* "Such isolated cases cannot establish linkage".

Cockayne, Penrose, Bell and Haldane, all members of the MRC Committee, no doubt had some interesting exchanges about Davenport's pedigree. Penrose was being unduly pessimistic – an isolated pedigree is sufficient so long as it is big enough and, as Bell and Haldane concluded, "The linkage here investigated is so close that on quite a small amount of material it has been possible to demonstrate its existence without leaving grounds for reasonable doubt".

The Galton Professorship

After the initiatives taken by Hurst and Hogben, the next important event was the election of R.A. Fisher to the Galton Professorship of Eugenics in 1933. Fisher had raised the possibility – or "contingency" as he called it – in a letter to Leonard Darwin in February 1929, saying "It would be easy to continue mathematical researches, and possibly in time to build up a reasonable biological outlook".

Early in 1933 Karl Pearson announced that he would retire at the end of the academic year. His department was split into the Galton Laboratory under Fisher as Galton Professor and a statistics department with Egon Pearson, Karl's distinguished statistician son, as Reader and head. The story of the resulting problems has often been told.

Haldane moved from Cambridge to University College London in the same year to become the first Professor of Genetics, also as part of the reorganisation consequent upon the elder Pearson's retirement. He had supported Fisher's election, writing to him "Please do not thank me in connection with your appointment. When asked my advice I mentioned a number of arguments against you, some of which were new to members of the committee. It was the merest regard for truth, and not any personal regard which I may feel for you, which forced me to add that you were the only possible candidate for the post". In 1936 Haldane was elected to the Weldon Professorship of Biometry at University College, established under Mrs. Weldon's will to promote "the higher statistical study of biological problems". It was held after Haldane by C.A.B. Smith, also of linkage fame.

The difficulties Fisher faced in trying to build up his department were astonishing, especially when viewed from our age in which money from the government and the medical charities seems to rain down on universities so abundantly that flooding is a major problem. He had soon secured the services of K. Mather, W.L. Stevens and Mrs. Sarah Holt (then Miss North) and had persuaded the Eugenics Society to give financial support to the *Annals of Eugenics*, the leading human genetics journal (now the *Annals of Human Genetics*), whose editorship went with the Galton professorship. From 1934 to 1940 the *Annals* was jointly published by the Galton Laboratory and the Eugenics Society.

The turning point in the fortunes of the Laboratory came with the establishment of the Serological Unit in 1935, funded by the Rockefeller Foundation. Already, in 1930, Fisher's interest in serology had been reinforced when Haldane told him of the work of Dr. Charles Todd with poultry, supported by the MRC. Then, "In the autumn of 1934", writes Joan Box, "Dr D.P. O'Brien came to England as the representative of the Rockefeller Foundation to consult with the Medical Research Council how best the foundation might sponsor research into human genetics. ... Among the members of the MRC's Committee on Human Heredity he met Fisher", and as a result of a subsequent meeting between them and the Provost of University College, Fisher wrote a research proposal for O'Brien. He asked for support for a "unit devoted to serological studies of accessible pedigrees of medical interest". The proposal was duly accepted, and the Serological Unit was funded from April 1935, with Dr. G.L. Taylor in charge. Others soon joined with MRC assistance, notably R.R. Race in 1937.

Linkage and Prognosis

Bell, Fisher, Haldane, Hogben and Penrose were all in London, all members of the MRC Committee and all sinking undoubted political differences to discuss linkage, its value, and its estimation. If the suggestion that the human genome could be mapped came from Bernstein via Hogben, then who suggested the very specific idea that the linkage of a disease locus with a blood-group marker might aid prognosis in the case of a late-onset disease? According to McKusick (1996), "Haldane suggested in the 1920s that diagnosis by the linkage principle would be both possible and useful". In a 1956 letter to *The Lancet* Professor Edwards made a similar suggestion for prenatal diagnosis using amniocentesis, but without any historical references.

My own first sightings of the proposal were for a long time in two papers of Fisher's in 1935. One of these is striking for three reasons: its title, its forum, and its omission from Fisher's *Collected Papers*. This short paper is entitled "Linkage studies and the prognosis of hereditary ailments" and was given to the International Congress on Life Assurance Medicine in London. I am fortunate to possess an offprint, culled from Fisher's offprint boxes in 1958 with his permission. Stating the reason for an interest in linkage, Fisher says "In this note I will give a brief account of the available methods for detecting linkage in man", which he does with great clarity, referring to Bernstein's original method, Haldane's improvement of it and Penrose's brand-new sib-pair method.

The second paper is particularly valuable because it is a kind of manifesto for the Galton Laboratory work and no doubt relies extensively on Fisher's submission to the Rockefeller Foundation the previous winter. Although Fisher was never a Galton Lecturer of the Eugenics Society, the paper is an address to the Annual General Meeting of the Society on 14 May 1935 entitled "Eugenics, academic and practical", a title apparently chosen for him. After some preliminaries, Fisher first congratulates the Society on its recent decision to establish Leonard Darwin Studentships. These were an initiative of Fisher's, and in a letter amongst his papers in Adelaide he lists the departments at which he considered they might appropriately be tenable, including both Hogben's and Haldane's, though one wonders what these two colleagues would have thought of housing students supported by the Eugenics Society. The studentships were continued after the war, and I held one myself in Fisher's old department in Cambridge in 1960–61.

Fisher goes on to emphasise the need for research in human genetics, mentioning first the importance of quantitative inheritance and then of dominance. Next, he describes Bell's continuing efforts under MRC auspices to obtain genealogical information on familial disorders, mentioning, in particular, her material on Huntington's chorea just published in the *Treasury of Human Inheritance*. At this point he refers to "another peculiarity brought to light by genetical research, namely linkage, which is likely in the future to revolutionise the methods of individual prognosis". "The search for such linkage will certainly be lengthy, and at first, disappointing". Calling the blood groups and markers such as the ability to taste PTC "harmless traits", Fisher writes, "But suppose we knew that one or other of these harmless traits which I have mentioned were closely linked in inheritance with Huntington's chorea, such knowledge might altogether change the situation [i.e. the calculation of probabilities]".

Fisher then acknowledges that "through the munificence of a great American foundation, we shall be able ... to establish a laboratory for serological genetics", and he explains the discoveries of new blood groups which he anticipates and how these will provide the markers by which to triangulate the genome. He closes his talk by warning against "crankery" and "the self-advertisement of irresponsible monomaniacs".

No sooner had I written the above than it occurred to me that perhaps the prognostical value of linkage might have been suggested by Haldane in 1923 in his famous speculative essay *Daedalus or Science and the Future* and that this might have generated McKusick's remark. Fortunately, I possess the 1995 reprint edited by Krishna Dronamraju where Professor Sir David Weatherall's chapter finally gives the clue, not to *Daedalus* itself, but to one of Haldane's essays in *Possible Worlds* published in 1927. I had bought a copy of *Possible Worlds* second-hand for 2/6 as an undergraduate, and that is indeed where we have all read the germ of the idea – and most of us have forgotten our source. It is fair to add that Haldane's speculation was not in connection with disease, but was a general observation that knowledge of the human genome would enable predictions to be made from marker genotypes about characters determined by loci linked to them – "landmarks for the study of such characters as musical ability, obesity and bad temper". "When that day comes intelligent people will certainly consider their future spouses' hereditary make up". "It is as certain that voluntary adoption of this kind of eugenics will come, as it is doubtful that the world will be converted into a human stud-farm". Whether Bernstein was familiar with *Possible Worlds* we will surely never know, but I expect Hogben was.

A search of Haldane's writings reveals something a little more specific in his Norman Lockyer Lecture *Human Biology and Politics* delivered in London on 28 November 1934. Perhaps Fisher was present. "... If we possessed the same knowledge of human genetics as we do of the genetics of *Drosophila* or maize, we should be able to say, with very high probability, that such and such children of a sufferer from Huntingdon's (*sic*) chorea has received a gene for it, and should not marry". So probably the safest conclusion is only that the idea arose in the Senior Common Room of University College presumably frequented by both Fisher and Haldane.

Conclusion

When Fisher wrote a foreword to Race and Sanger's *Blood Groups in Man* in 1950 he referred to "that 'basic triangulation' by which in due time the whole [human germ plasm] will be surveyed". Fifty years on the initial survey is complete, but the triangulation itself, the linkage map, still has a long way to go. The achievement of the 1920s and 1930s was to establish Mendelian human genetics on a firm base separate from the wild enthusiasms of the earlier years, and to provide it with a calculus for linkage analysis whose descendants are recognisable today.

Sources

I am much indebted to two books, Kevles's *In the Name of Eugenics* already mentioned and Pauline Mazumdar's *Eugenics, Human Genetics, and Human Failings* (1992). There are biographies of three of the main participants: *J.B.S.: The Life and Work of J.B.S. Haldane*, by Ronald Clark (1968); *R.A. Fisher: The Life of a Scientist*, by his daughter Joan Fisher Box (1978); and *Lancelot Hogben, Scientific Humanist: An Unauthorised Autobiography*, edited by Adrian and Anne Hogben (1998). Nor should we forget the books which these men themselves published. Haldane's *New Paths in Genetics* did not appear until 1941, but Hogben published *Genetic Principles in Medicine and Social Science* in 1931 and *Nature and Nurture: the William Withering Lectures on the Method of Clinical Genetics* in 1933. Both men were, in addition, prolific essayists. Fisher's *The Genetical Theory of Natural Selection* has already been mentioned. Not such an essayist as Haldane and Hogben, Fisher nevertheless wrote an astonishing number of reviews and annotations for the *Eugenics Review* – with the list in his biography consists of four pages of small type – the major portion appearing between 1916 and 1935. His correspondence with Leonard Darwin and others has been edited by J.H. Bennett (1983). There is a short and superficial biography of Penrose by Smith (2001) and a full account of his attitude to eugenics in two papers by Watt (1998).

The Fisher papers are in the Barr Smith Library of the University of Adelaide and the Hurst papers are in Cambridge University Library. Many of the scientists mentioned were Fellows of the Royal Society of London and thus the subjects of extensive biographical notices, always very valuable through being written by experts.

For two personal accounts of later developments in linkage estimation see C.A.B. Smith (1986) and Newton Morton (1995). The German contribution needs its own historian – a brief introduction is provided by Crow (1993): "Felix Bernstein and the first human marker locus", whilst Hogben (1931, 1933) described Bernstein's linkage method. J.H. Edwards (1993) has described Haldane's contribution, and Part IV of N.T.J. Bailey's *Introduction to the Mathematical Theory of Genetic Linkage* (1961) is devoted to human genetics. There is a substantial correspondence between Fisher and W. Weinberg from the 1930s in the Fisher archive at the University of Adelaide. Fisher also corresponded with Bernstein, whom he honoured by having him elected to an Honorary Fellowship of the Royal Statistical Society during his own presidency.

References

Bailey, N.T.J. (1961) *Introduction to the Mathematical Theory of Genetic Linkage*. Oxford: Clarendon Press.

Bell, J. and Haldane, J.B.S. (1937) The linkage between the genes for colour-blindness and haemophilia in man. *Proc. Royal Society of London B* **123**, 119–50.

Bennett, J.H., ed. (1983) *Natural Selection, Heredity, and Eugenics. Including selected correspondence of R.A. Fisher with Leonard Darwin and others*. Oxford: Clarendon Press.

Box, J.F. (1978) *R.A. Fisher: The Life of a Scientist*. New York: Wiley.

Clark, R. (1968) *J.B.S.: The Life and Work of J.B.S. Haldane* London: Hodder and Stoughton.

Cockayne, E.A. (1933) *Inherited Abnormalities of the Skin and its Appendages*. London: Oxford University Press.

Crow, J.F. (1993) Felix Bernstein and the first human marker locus. *Genetics* [January]. Reprinted in Crow and Dove (2000), 320–3.

Crow, J.F. and Neel, J.V., eds. (1967) *Proceedings of the Third International Congress of Human Genetics*. Baltimore: Johns Hopkins.

Crow, J.F. and Dove, W.F., eds. (2000) *Perspectives on Genetics*. Madison: University of Wisconsin Press.

Davenport, C.B. (1930) Sex linkage in man. *Genetics* **15**, 401–44.

Dronamraju, K.R., ed. (1968) *Haldane and Modern Biology*. Baltimore: Johns Hopkins.

Dronamraju, K.R. (1985) *Haldane*. Aberdeen University Press.

Dronamraju, K.R., ed. (1995) *Haldane's Daedalus Revisited*. London: Oxford University Press.

Edwards, A.W.F. (1972) *Likelihood*. Cambridge: Cambridge University Press; Expanded edition Johns Hopkins University Press, 1992.

Edwards, A.W.F. (1997) The early history of the statistical estimation of linkage. *Annals of Human Genetics* **60**, 237-249.

Edwards, A.W.F. (2000) *The Genetical Theory of Natural Selection*. In *Perspectives*, ed. J.F. Crow and W.F. Dove. *Genetics* **154**, 1419–26.

Edwards, A.W.F. (2001) Darwin and Mendel united: the contributions of Fisher, Haldane and Wright up to 1932. In *Encyclopedia of Genetics*, ed. E.C.R. Reeve, London: Fitzroy Dearborn, 77–83.

Edwards, J.H. (1956) Antenatal detection of hereditary disorders. *Lancet* **270**, 579.

Edwards, J.H. (1969) The application of genetics to man. In *Fifty Years of Genetics*, ed. J Jinks, Edinburgh: Oliver and Boyd, 67–79.

Edwards, J.H. (1993) Haldane and the analysis of linkage. In *Human Population Genetics*, ed. P.P. Majumder; New York: Plenum, 153–64.

Fisher, R.A. (1918) The correlation between relatives on the supposition of Mendelian inheritance. *Transactions of the Royal Society of Edinburgh* **52**, 399-433.

Fisher, R.A. (1930) *The Genetical Theory of Natural Selection.* Oxford: Clarendon Press; 2nd ed: New York: Dover Publications, 1958; variorum ed: London: Oxford University Press, 1999.

Fisher, R.A. (1935) Linkage studies and the prognosis of hereditary ailments. *Transactions of the International Congress on Life Assurance Medicine,* London.

Fisher, R.A. (1935) Eugenics, academic and practical. *Eugenics Review* **27**, 95–100.

Fisher, R.A. (1971-74) *Collected Papers of R.A. Fisher,* Vols 1-5, ed. J.H. Bennett, University of Adelaide.

Haldane, J.B.S. (1924) "Daedalus, or, Science and the Future", a paper read to the Heretics, Cambridge, on February 4.

Haldane, J.B.S. (1927) *Possible Worlds.* London: Chatto and Windus.

Haldane, J.B.S. (1934) *Human Biology and Politics.* London: British Science Guild.

Haldane, J.B.S. (1938) *Heredity and Politics.* London: Allen and Unwin.

Haldane, J.B.S. (1941) *New Paths in Genetics.* London: Allen and Unwin.

Haldane, J.B.S. and Moshinsky, P. (1939) Inbreeding in Mendelian populations with special reference to human cousin marriage. *Annals of Eugenics* **9**, 321–40.

Hogben, L. (1931) *Genetic Principles in Medicine and Social Science.* London: Williams and Norgate.

Hogben, L. (1933) *Nature and Nurture.* London: Allen and Unwin.

Hogben, L. (1998) *Lancelot Hogben, Scientific Humanist.* Rendlesham, Suffolk: Merlin.

Hogben, L. (1998) *Lancelot Hogben, Scientific Humanist: An Unauthorised Autobigraphy.* Edited by Adrain and Anne Hogben. Rendlesham, Suffolk: Merlin.

Hurst, C.C. (1932) *The Mechanism of Creative Evolution.* Cambridge, UK: Cambridge University Press.

Kevles, D.J. (1985) *In the Name of Eugenics: Genetics and the Uses of Human Heredity.* New York: Knopf.

Keynes, M., ed. (1993) *Sir Francis Galton, F.R.S.* London: Macmillan.

McKusick, V.A. (1996) History of medical genetics. In *Emery-Rimoin Principles and Practice of Medical Genetics,* 3rd ed., eds. Rimoin, D.L., Connor, J.M. and Pyeritz, R.E.; Edinburgh: Churchill Livingstone, 1–30.

Mazumdar, P.M.H. (1992) *Eugenics, Human Genetics, and Human Failings.* London: Routledge.

Moran, P.A.P. and Smith, C.A.B. (1966) Commentary on R.A. Fisher's paper on "The correlation between relatives on the supposition of Mendelian inheritance". Eugenics Laboratory Memoirs XLI, Galton Laboratory, London.

Morton, N.E. (1995) LODs past and present. *Genetics* [May]. Reprinted in Crow and Dove (2000), 453–8.

Mourant, A.E. (1954) *The Distribution of the Human Blood Groups.* Oxford: Blackwell.

Pearl, R. (1914) Studies in inbreeding V. *American Naturalist* **48**, 513–23.

Penrose, L.S. (1934) *The Influence of Heredity on Disease.* London: Lewis.

Penrose, L.S. (1967) The influence of the English tradition in human genetics. *Proceedings of the Third International Congress of Human Genetics,* Baltimore: Johns Hopkins, 13–25.

Provine, W.N. (1971) *The Origins of Theoretical Population Genetics.* Chicago: University of Chicago Press.

Punnett, R.C. (1950) Early days of genetics. *Heredity* **4**, 1–10.

Race, R.R. and Sanger, R. (1950) *Blood Groups in Man.* Oxford: Blackwell.

Smith, C.A.B. (1986) The development of human linkage analysis. *Annals of Human Genetics* **50**, 293–311.

Smith, M. (2001) *Lionel Sharples Penrose: A Biography.* Colchester: Michael Smith.

Watt, D.C. (1998) Lionel Penrose, F.R.S. (1898–1972) and Eugenics. *Notes and Records of the Royal Society of London* **52**, 137–51 (Part One), 339–54 (Part Two).

5. William Bateson, Archibald Garrod and the Nature of the "Inborn"[1]

Patrick Bateson

William Bateson coined the term "genetics" and was the most vigorous promoter of Mendel at the beginning of the twentieth century. His contemporary, Archibald Garrod, first used the term "inborn errors of metabolism" and may be justifiably regarded as the founder of biochemical genetics, which has had such an important role in contemporary medicine. The linking of Bateson and Garrod not only brings together two people whose influence over the next 100 years was enormous, but also links molecular biology together with the study of the whole organism, thereby providing a bridge between the mechanisms of gene expression and the principles of biological inheritance.

I am not geneticist myself, nor am I medically qualified. I study the biology of behaviour and have a particular interest in how behaviour develops. With such an interest, I have had to wrestle with nature-nurture issues throughout my professional life, and the question of what is inborn has proved endlessly teasing and challenging. While I am not directly descended from William Bateson, he was the cousin of my grandfather, and this relationship probably explains why, from a very early age, I told the world that I wanted to be a biologist without having any clear idea what that might entail. The two sides of the family had a close link because, when I was a boy, my parents cared for Will's younger brother Ned when he was a widower and a very old man. Everybody remarked on the astonishing resemblance between Ned and myself and, as I subsequently discovered years later, I also looked very much like William's son Gregory, even though he was only a second cousin once removed. I shall return to family likenesses later.

Bateson and Garrod were born within four years of each other, Garrod on 25 November 1857 and Bateson on 8 August 1861. They were both raised in comfortable homes and both had eminent fathers – archetypal Victorian intellectuals. Bateson's father was Master of St. John's College, Cambridge for 24 years and Garrod's father was a distinguished physician, subsequently knighted, and a good friend of Francis Galton. Both sons obtained firsts in Natural Sciences, Bateson specialising in Zoology in Cambridge and Garrod specialising in Chemistry in Oxford. Both went on to have famous careers themselves and both became Fellows of the Royal Society; Garrod was knighted in 1918 and Bateson, the less conventional of the two, turned down a knighthood in 1922. Perhaps most important of all for my story, the careers of the two men intersected at a crucial stage for both of them. Finally, both had ideas about biological evolution which were largely ignored for many years.

---

[1] The Darwin Lecture 2001

Bateson did his first major study on the embryology of an animal called *Balanoglossus* living on tidal mudflats; it looks like a worm, but is now regarded as a primitive vertebrate. It was particularly common in Chesapeake Bay on the East Coast of the United States and it was there that Bateson got to know a brilliant American zoologist, William K. Brooks, who was bringing out a book about heredity at the time (Brooks, 1883). Although few biologists doubted that Charles Darwin had provided the most coherent and complete explanation for adaptation by the process of natural selection, the necessary conditions for one species to become distinct from another remained a source of dispute. Darwin's mechanism for evolutionary change consisted of three crucial steps. Each step must have been in place if adaptation by the organism to the environment occurred in the course of biological evolution. First, variation must have existed. Second, some variants must have survived more readily than others. Third, the variation must have been inherited.

While Darwin's proposal provided a powerful and plausible mechanism for generating adaptations, it was less obvious that it would provide what was needed for the formation of a new species. His friends, Thomas Henry Huxley and Francis Galton, were doubtful and so, it turned out, was Brooks. Much of this scepticism rubbed off on Bateson in the course of their long discussions after work in Chesapeake Bay.

Bateson was sometimes thought to be anti-Darwinian (Bowler, 1983), but such a view totally misconstrues what he was after. In the first phase of his life's work, he wanted to know what variation within a species might look like. He was particularly interested in finding major discontinuities in characters. He amassed a great quantity of examples in his book *Materials for the Study of Variation* which appeared in 1894 (Bateson, W., 1894). He believed that such discontinuities could provide evidence for steps that might lead to the appearance of a new species. This interest prepared him for the rediscovery of Mendel's work, which provided the rules for how the qualitative differences between members of the same species could be inherited. He was not opposed to the Darwinian proposal for evolutionary change, but did not share Darwin's belief that evolution of new species had always involved continuous modification.

Having established that discontinuities were to be found in nature, the next step was to discover what happened to such discontinuities from one generation to the next. Bateson set to work on the experimental breeding of animals and plants in order to find out how that variation might be inherited. Before the rediscovery of Mendel's work, the best known principle was the Law of Ancestral Heredity, promulgated in the nineteenth century most actively by Francis Galton (1897). Those of a mathematical bent liked it because it meant that a prediction about the characteristics of individuals could be derived probabilistically from the characteristics of the ancestors. Galton produced two forms of the Law of Ancestral Heredity, but these were mathematically inconsistent (Provine, 1971). Karl Pearson subsequently cleaned up the mathematics (Pearson, 1898) but did not help matters when he called his revision "Galton's theory". Confusion reigned. The independent rediscovery of Mendel's work by de Vries, Tschermak and Correns changed all that (an enjoyable account is given by Henig, 2000). When Bateson became aware of Mendel's work he realised in a flash that here were the principles that would make sense of heredity.

Mendel's discoveries may be summarised as follows. Inherited factors influencing the characteristics of an organism come in pairs; while this is usually true, sex linkage (arising from unequal pairings of the sex chromosome in male mammals and female birds) was only discovered later. One factor is often dominant to the other and the recessive one lies latent within an individual. It is a matter of chance which of each paired factors enters into the gamete that fuses with the gamete of the other parent. The particular factor that enters a gamete is usually unrelated to the member of another pair of factors; the linkage of some factors was only to emerge later.

In 1902 Bateson renewed an intellectual battle with the biometrician Walter Frank Raphael Weldon. This bitter struggle had started when his former friend and mentor wrote a critical review of *Materials for the Study of Variation* eight years earlier. Weldon now tried to argue that Mendel simply described a special case and, in any event, the results could be explained in terms of the Law of Ancestral Heredity. This patronising comment fired up Bateson who, with his enormous energy and determination, wrote in a few months a fierce rebuttal of Weldon's review in the book called *Mendel's Principles of Heredity: A Defence* (Bateson, W., 1902). Weldon had further infuriated Bateson by concluding his review of Mendel's paper by writing that, without wishing to belittle Mendel's achievement, he wanted "to call attention to a series of facts which seem to me to suggest fruitful lines of enquiry". Bateson commented that Weldon was about as likely to kindle interest in Mendel's discoveries as to light a fire with a wet dishcloth.

When Weldon suddenly died in 1906, Bateson wrote to his wife Beatrice of this fierce squabble with his former friend: "If any man ever set himself to destroy another man's work, that did he do to me …", but in another letter he wrote: "To Weldon I owe the chief awakening of my life. It was through him that I first learnt that there was work in the world which I could do. … Such a debt is perhaps the greatest that one man can feel towards another …" (Bateson, B., 1928). Shaken by Weldon's death, Bateson offered an olive branch to Karl Pearson, the principal biometrician of the time, but the peace offering was rejected and the battle between the Mendelians and the biometricians persisted.

Bateson's commonsense rejection of the biometricians' premature attempts to formalise the principles of heredity does seem justified now. R.A. Fisher believed that "… had any thinker in the middle of the nineteenth century undertaken, as a piece of abstract and theoretical analysis, the task of constructing a particulate theory of inheritance, he would have been led, on the basis of a few very simple assumptions, to produce a system identical with the modern scheme of Mendelian or factorial inheritance" (Fisher, 1930) p. 7. But nobody had been led to the deduction that inherited factors influencing the characteristics of an organism come in pairs – or at least that they usually do. No amount of clever mathematics could have led to the deduction that one is often dominant to the other. (Galton understood that many "gemmules", as

Darwin had called them, capable of influencing the characteristics of an organism, must often lie latent, but that wasn't a mathematical deduction – it was based on empirical observation.) No amount of clever mathematics could have led to the deduction that it is a matter of chance which of each paired factors enters into the gamete that fuses with the gamete of the other parent. And no amount of clever mathematics could have led to the deduction that the particular factor that enters a gamete is usually unrelated to the member of another pair of factors. Once known, the stage would be set for formalisation - but not before.

A mathematician, Udny Yule (1902), pointed out at an early stage that the struggle between Bateson and the biometricians was entirely unnecessary since Mendelian factors could give rise to small changes and therefore be compatible with Darwin's view that evolutionary change was continuous. The differences between the Mendelians and the biometricians was primarily over whether discontinuous change could occur. Evidence of discontinuity, provided by "sports" – strikingly different phenotypes – was regarded by the Mendelians as evidence against Darwin. When R.A. Fisher finally demonstrated to universal satisfaction that Mendelism could be reconciled with Darwin's notion of continuous evolutionary change, the source of the controversy seemed to have been removed (Fisher, 1930). Fisher, Sewall Wright and J.B.S. Haldane brought mathematical rigour to the subject and founded the field of theoretical population genetics. While they agreed upon the importance of Darwinian evolution, each of them produced a distinct model and some inconsistencies between the theoretical frameworks on which their subject is based remain to this day. From the standpoint of Darwinian theory, Ernst Mayr (1942) persuasively argued that small isolated populations could rapidly evolve distinct characteristics that made them genetically incompatible with closely related populations, thus forming a new species. From this perspective, it might seem in hindsight that Bateson's search for discontinuities in order to explain the origin of species was a waste of time. However, the debate is far from over and, indeed, a view is growing that Bateson has been unjustly maligned (see Forsdyke, 2001).

Sudden discontinuities in evolution have been given prominence by modern palaeontologists who have been impressed by periods of stasis and sudden change in the fossil record (Eldredge, 1995; Gould, 2002). They suggest that, after periods of stasis in evolution, sudden changes can occur in the fossil record and these may represent the appearance of new species. This idea has recurred periodically and, notably, was foreshadowed in the writings of Goldschmidt (1940) who, in a memorable phrase, referred to a fresh arrival, that might give rise to a new species, as a "Hopeful Monster". Galton (1892) had already produced a vivid image of how Darwin's Law of Continuity might be satisfied by a series of changes in jerks.

> The mechanical conception would be that of a rough stone, having, in consequence of its roughness, a vast number of natural facets, on any one of which it might rest in "stable" equilibrium. That is to say, when pushed it would somewhat yield, when pushed much harder it would again yield, but in a less degree; in either case, on the pressure being

withdrawn it would fall back into its first position. But, if by a powerful effort the stone is compelled to overpass the limits of the facet on which it has hitherto found rest, it will tumble over into a new position of stability, whence just the same proceedings must be gone through as before, before it can be dislodged and rolled another step onwards (pp. 354-355).

Hosts of examples of big events having no effect and small events leading to big changes are to be found and many of these are now formalised by the non-linear mathematical techniques derived from Catastrophe Theory and Chaos.

Hopeful Monsters were disparaged on the grounds that even if a big change in the phenotype could occur as a result of mutation, the Hopeful Monster would be a novelty on its own with no possibility of finding a mate. Without a mate there would be no new species. However, if we suppose that, somehow or other, there were enough Hopeful Monsters to breed successfully with each other, the possibility exists of competition between the Hopeful Monsters and the stock from which they sprang. It is not at all difficult to suppose that, by the process of natural selection, Hopeful Monsters could quickly replace their competitors if they were better adapted to the environment. No new fancy principles of evolution are involved here. One set of requirements for a sudden change in evolution is that: (a) a large change in phenotypic expression arises as the result of small underlying changes; (b) the new phenotype is sufficiently frequent in the population to allow breeding to occur; and (c) the new phenotype is more successful that the old. Such requirements could be met in many ways (see Bateson, P., 1984).

The battle over discontinuous variation held up moves towards mathematical formalisations of genetics for many years. In his superb history of population genetics, William Provine (1971) felt that if, after Weldon's death, Bateson and Pearson had collaborated instead of fought, population genetics would have gained a significantly earlier start. Science is created by real people and argumentative, uncompromising Bateson was real enough for anybody. But if he had been bland, Mendel would probably have remained unchampioned and might well have been disregarded. In all likelihood, the link between the two heroes of my story would never have been forged.

The link came at the very beginning of the twentieth century. By this stage Garrod was medically qualified and a well-established physician at Great Ormond Street. After leaving school, Garrod had been sent to Christ Church, Oxford by his distinguished father who feared that Archie might be outshone by his brilliant brother, Alfred Henry, who already had won a research fellowship at St. John's College, Cambridge. Despite his father's fears, Archie did extremely well in Chemistry at Oxford and this training stood him in very good stead in his subsequent medical career. He went to St. Bartholomew's Hospital and qualified in medicine in 1885, receiving a Doctorate of Medicine from Oxford in 1886. He spent a year travelling and was especially influenced by his experiences in the Allgemeines Krankenhaus in Vienna (Bearn, 1993). He spent the early part of his career becoming a well-rounded physician with eight years at the West London Hospital, as well as at Bart's from 1887, where he became Assistant Physician in 1903 and Full Physician in 1912. At the time of his appointment as Assistant Physician to the Hospital for Sick Children, Great Ormonde Street in 1892 (Full Physician in 1899) he was already showing signs of becoming interested in the hereditary basis of disease.

Garrod had become particularly interested in a rare abnormality in which a person produces urine that blackens on exposure to air. It is highly noticeable at an early stage in life because the babies' nappies are stained deeply by the black urine. The critical compound in the urine of people with alkaptonuria, as the condition is called, is homogentisic acid of which 2.5-6.0 grams is produced each day.

Alkaptonuria is much more common in men than women and, though always rare, is particularly likely to occur in the offspring of first cousin marriages. Bateson got to hear of this and in December 1901 he and Saunders reported Garrod's finding to the Evolution Committee of the Royal Society. They wrote "... the mating of first cousins gives exactly the conditions most likely to enable a rare and usually recessive character to show itself. If the bearer of such a gamete mates with individuals not bearing it, the character would hardly ever be seen; but first cousins will frequently be bearers of *similar* gametes, which may in such unions meet each other, and thus lead to the manifestation of the peculiar recessive characters in the zygote" (Evolution Committee of the Royal Society, 1902).

Alexander Bearn (1993), biographer of Garrod, found a long letter from Garrod to Bateson dated 11 January 1902 beginning: "It was a great pleasure to receive your letter and to learn that you are interested in the family occurrence of alkaptonuria." This suggests that Bateson initiated the correspondence. Anyway, Garrod was quick to see the significance of Mendelism for congenital human conditions and referred explicitly to this insight in his next *Lancet* paper (Garrod, 1902). No case of two alkaptonurics having children together was known. Garrod had both male and female alkaptonuric patients and he used to get them into the ward at the same time in the hope that they might become fond of each other, marry and have children (Bearn, 1993). If the children had been alkaptonuric, that would have clinched the Mendelian hypothesis.

In his 1902 *Lancet* paper Garrod suggests that alkaptonuria is a harmless alternative mode of metabolism which seems to have no adverse effect on the health of the person with the condition. Nobody had doubted that individuals looked different from each other but, as Bearn points out, it was an entirely new and far-reaching concept to suppose that each person's chemical make-up was individually distinct.

Notwithstanding the view that each person's biochemistry might be unique, Garrod was subsequently to coin his famous phrase "inborn *errors* of metabolism". Of course, the mutation of a gene affecting the structure and function of an enzyme could easily be an error, in biological terms, and carry with it important medical consequences for the person. Even alkaptonuria leads to a disease of the cartilage and patients may develop joint pains and severe backache (T.M. Cox, 2000). Anyhow, over the years many examples of diseases were found in which the medical consequences were severe (Harris, 1963). One of the most famous examples of such a disease is phenylketonuria. Here the enzyme, phenylalanine hydoxylase, which normally catalyses the conversion of phenylalanine to tyrosine is deficient (Jervis, 1953). Phenylalanine builds up and is converted into other products, some of which are toxic and have a variety of non-specific effects on the person. As adults, these people have small brains, abnormal brain rhythms, make repetitive and unusual movements of the hands, rock rhythmically for hours and have serious cognitive disabilities; to cap it all for their unfortunate parents, they have severe temper tantrums. All of these pathological effects of the inborn error can be prevented by feeding the child a diet in which the presence of phenylalanine has been restricted.

Garrod's linking of inborn errors of metabolism to genetics was the first step in a historical process that has had enormous implications for medicine through the pharmaceutical industry and the highly promising possibilities for modelling disease in animals by genetic modification (Bateson, P., et al., 2001). Since the primary biochemical changes are usually linked closely to genetic mutations, the temptation to think of genetic blueprints or codes for biochemical characteristics of the whole organism is very strong. In such cases, there is, indeed, likely to be a one-to-one correspondence between gene and protein. It was natural that Garrod's pioneering efforts should lead to the one gene-one enzyme hypothesis (Beadle, 1945). Of course, many proteins are not enzymes and play a structural, a messenger or another non-catalytic role in the construction and maintenance of the body. Nobody can gainsay the historical or the practical importance of Garrod's achievement. But the cost of such relentless reductionism has also been considerable, because the metaphors of blueprints or codes become seriously misleading when analysis moves to the level of functional systems or whole organisms. A blueprint implies a one-to-one correspondence between the plan and the building. That is not what we find in biology as soon as we move away from biochemistry – and even when we stay within it.

Bateson was obviously keenly aware of the interactions between genes. He did a famous experiment with Punnett in which he crossed two white strains of chicken, White Silkies with White Dorkings. The offspring were not white. They were coloured, so it became clear that something came from a White Silky parent that interacted with something else that came from the White Dorking and it was this interaction that produced the colour. Even the archetypal case for inborn errors of metabolism suggested an interaction. Alkaptonuria was reported as being much more common in men than women (Garrod, 1902). If this were not simply that men were more likely to bring their condition to the attention of a medical doctor, the sex difference would indicate that the expression of the recessive gene was affected by another gene on the Y chromosome.

Of the three great figures who started the formalisation of population genetics, Sewall Wright was the most sensitive to the interactions between genes (Wright, 1930). Sewall Wright believed that selection for single genes was far less effective than the selection of interaction systems. In this way he was much more like Bateson than either of the other two great architects of theoretical population genetics. Fisher was keen to isolate the non-additive effects in his equations so that he could deal with the much more tractable additive effects.

However, the mathematical brilliance has arguably got in the way of understanding the biological phenomena. The point can be made by looking at the details of a modern example provided by people with the Kallmann syndrome.

The main behavioural consequence of the Kallmann syndrome in men is a lack of sexual interest in members of either sex. The syndrome is caused by damage at a specific genetic locus (Pfaff, 1997). The syndrome was classically described as sex linked, but other genes that have been found to produce the same syndrome are autosomal. Cells that are specialised to produce a chemical messenger called gonadotropin-releasing hormone (GnRH) are formed initially in the nose region of the foetus. Normally the hormone-producing cells would migrate into the brain. As a result of the genetic defect, however, their surface properties are changed and the cells remain dammed up in the nose. The activated GnRH cells, not being in the right place, do not deliver their hormone to the pituitary gland at the base of the brain. Without this hormonal stimulation, the pituitary gland does not produce the normal levels of two other chemical messengers, luteinizing hormone and follicle-stimulating hormone. Without these hormones, the testes do not produce normal levels of the male hormone testosterone. Without normal levels of testosterone, the man shows little sign of normal adult male sexual behaviour. Even in this relatively straightforward example, the pathway from gene to behaviour is long, complicated and indirect. Each step along the causal pathway requires the products of many genes and has ramifying effects, some of which may be apparent and some not.

This brings me to the issue that relates most strongly to my own area of expertise. Here the prolonged discussions in my own field of ethology have been especially helpful. A range of examples encourage the view that even an aspect of biology as complex as behaviour can be inborn.

Very simple rules for responding to chemical gradients can underlie seemingly purposive behaviour of protozoa. Simple feedback mechanisms can explain the aggregation of wood lice in damp places. Even highly flexible behaviour is amenable to approaches that look for simplicity behind the complexity. A spider building an orb web explores a potential site, creates a frame of web round the various attachment points, spins radials from the attachment points to the orb, spins more radials to the frame, then spins a spiral from the orb to the outside and finally spins another spiral from the frame to the orb. It is an exquisite structure adjusted to the site in which it is built. It looks complicated and the construction of the web is possible across a wide range of environmental conditions. Very simple rules can, nevertheless, be devised to simulate exactly what the spider does (Vollrath, Downes and Krackow, 1997). If the spider is controlled in the same way, the inheritance of what is needed and the development of the requisite rules need be no more problematic than building a kidney. A marvel but not a special marvel just because behaviour is involved.

Another example is the European garden warblers that have been hand-reared in cages, nevertheless, become restless and attempt to fly south in the autumn – the time when they would normally migrate southwards. The warblers continue to be restless in their cages for about a couple of months, the time taken to fly from Europe to their wintering grounds in Africa. The following spring, they attempt to fly north again. This migratory response occurs despite the fact that the birds have been reared in social isolation, with no opportunities to learn when to fly, where to fly or for how long (Gwinner, 1996).

Certain aspects of human behavioural development recur in everybody's life despite the shifting sands of cultural change and the unique contingencies of any one person's life. Despite the host of genetic and environmental influences that contribute to behavioural differences between individuals, all members of the same species are remarkably similar to each other in many aspects of their behaviour - at least, when compared with members of other species. All humans have the capacity to acquire language, and the vast majority do. With few exceptions, humans pass the same developmental milestones as they grow up. Most children have started to walk by about 18 months after birth, have started to talk by around two years and go on to reach sexual maturity before their late teens. Individual differences among humans seem small when any human is compared with any chimpanzee.

Human facial expressions have characteristics that are widely distributed in people of many different cultures. The emotions of disgust, fear, anger and pleasure are read off the face with ease in any part of the world. Towards the end of his life Charles Darwin (1872) wrote *The Expression of the Emotions in Man and Animals*, a book that provided the stimulus for observational studies of animal and human behaviour which have continued into modern times. Darwin concluded: "That the chief expressive actions, exhibited by man and by the lower animals, are now innate or inherited, – that is, have not been learnt by the individual, – is admitted by every one." Darwin's descriptions of suffering, anxiety, grief, joy, love, sulkiness, anger, disgust, surprise, fear and much else are models of acute observation. He would show to friends and colleagues pictures of people seemingly expressing various emotions and ask them, without further prompting, to describe the emotions. In one case a picture of an old man with raised eyebrows and open mouth was shown to 24 people without a word of explanation, and only one did not understand what was intended. In a way that shows both his carefulness and his honesty, Darwin continued: "A second person answered terror, which is not far wrong; some of the others, however, added to the words surprise or astonishment, the epithets, woeful, painful, or disgusted." His extensive correspondence with travellers and missionaries convinced him that humans from all over the globe expressed the same emotion in the same way. Subsequently, an enormous archive of photographic records of human expressions in different cultures at different stages of economic development was established (Ekman et al., 1987). The similarities in, for example, the appearance of the smile or the raised eyebrows are striking. The cross-cultural agreement in the interpretation of complex facial expressions is also remarkable. People agree about which emotions are being expressed. They also agree about which emotion is the more intense, such as which of two angry people seems the more angry (Ekman et al., 1987).

All of this might be taken to suggest that genes control behaviour just as they do biochemistry. However, the concept of the inborn as it is applied to the behaviour of whole organisms is riddled with confusion. It turns out that the term "instinct" means remarkably different things to different people. To some, "instinct" means a distinctly organised system of behaviour patterns, such as that involved in searching for and consuming food. For others, an instinct is simply behaviour that is not learned. Instinct has also been used as a label for behaviour that is present at birth (the strict meaning of "innate") or, like sexual behaviour, patterns that develop in full form at a particular stage in the life-cycle. Another connotation of instinct is that once such behaviour has developed, it does not change. Instinct has also been portrayed as behaviour that develops before it serves any biological function, like some aspects of sexual behaviour. Instinct is often seen as the product of Darwinian evolution so that, over many generations, the behaviour was adapted for its present use. Instinctive behaviour is supposedly shared by all members of the species (or at least by members of the same sex and age). Instinct has also been used to refer to a behavioural difference between individuals caused by a genetic difference (Bateson, P. and Martin, 1999).

One aspect of the unitary concept of instinct that unravelled on further inspection was the belief that learning does not influence such behaviour patterns once they have developed. Many cases of apparently unlearned behaviour patterns are subsequently modified by learning after they have been used for the first time. A newly hatched laughing gull chick will immediately peck at its parent's bill to initiate feeding, just as, in the laboratory, it will peck at a model of an adult's bill. At first sight this behaviour pattern seems to be unlearned; the chick has previously been inside the egg and therefore isolated from any relevant experience, so it cannot have learned the pecking response. However, as the chick profits from its experience after hatching, the accuracy of its pecking improves and the kinds of model bill-like objects which elicit the pecking response become increasingly restricted to what the chick has seen (Hailman, 1987). Here, then, is a behavioural response that is present at birth, species-typical, adaptive and unlearned, but nonetheless modified by the individual's subsequent experience.

Essentially the same is true for the "innate" smiling of a human baby. Human babies who have been born blind and, consequently, never been able to see a human face, nevertheless start to smile at around five weeks – the same age as normal babies. Babies do not have to see other people smile in order to smile themselves (Freedman, 1964). Just after birth, sighted human babies gaze preferentially at head-like shapes that have the eyes and mouth in the right places. Invert these images of heads, or jumble up the features, and the newborn babies respond much less strongly to them. Despite these observations, sighted people subsequently learn to modify their smiles according to their experience, producing subtly different smiles that are characteristic of their particular culture (Troster and Brambring, 1992). Nuance becomes important. The blind child, lacking the visual interaction with its mother, becomes less responsive and less varied in its facial expression. The fact that a blind baby starts to smile in the same way as a sighted baby does not mean that learning has no bearing on the later development of social smiling. Experience can and does modify what started out as apparently unlearned behaviour.

Conversely, some learned behaviour patterns are developmentally stable and virtually immune to subsequent modification. The songs of some birds are learned early in life, but these learned songs may be extremely resistant to change once they have been acquired.

The idea that one meaning of instinct, "unlearned", is synonymous with another, namely "adapted through evolution", also fails to stand up to scrutiny. The development of a behaviour pattern that has been adapted for a particular biological function during the course of the species' evolutionary history may nonetheless involve learning during the individual's lifespan. For example, the strong social attachment that young birds and mammals form to their mothers is clearly adaptive and has presumably evolved by Darwinian evolution. And yet the attachment process requires the young animal to learn the individual distinguishing features of its mother.

Yet another way in which the different elements of instinct fall apart is the role of learning in the inheritance of behaviour across generations. Consider, for example, the ability of birds such as titmice to peck open the foil tops of the milk bottles that used to be delivered each morning to the doors of a great many British homes. The birds' behaviour is clearly adaptive, in that exploiting a valuable source of fatty food undoubtedly increases the individual bird's chances of surviving the winter and breeding. However, the bottle-opening behaviour pattern is transmitted from one generation to the next by means of social learning. The basic tearing movements used in penetrating the foil bottle-top are also used in normal foraging behaviour and are probably inherited without learning. But the trick of applying these movements to opening milk bottles is acquired by each individual bird through watching other birds do it successfully – that is, by social learning. (How the original birds first discovered the trick is another matter.)

In short, many behaviour patterns have some, but not all, of the defining characteristics of instinct, and the unitary concept starts to break down under closer scrutiny. The various theoretical connotations of instinct – namely that it is unlearned, caused by a genetic difference, adapted over the course of evolution, unchanged throughout the life-span, shared by all members of the species, and so on – are not merely different ways of describing the same thing. Even if a behaviour pattern is found to have one diagnostic feature of instinct, it is certainly not safe to assume that it will have all the other features as well. It is worth remembering Darwin's wise reluctance to define "instinct": "An action, which we ourselves require experience to enable us to perform, when performed by an animal, more especially by a very young one, without experience, and when performed by many individuals in the same way, without their knowing for what purpose it is performed, is usually said to be instinctive. But I could show that none of these characters are universal. A little dose of judgment or reason ... often comes into play, even with animals low in the scale of nature" (Darwin, 1859).

Should the terminological confusion worry us? Not, in my view, if we adopt a Darwinian perspective. One of the triumphs of behavioural biology in the latter part of the twentieth century was to relate differences in mating systems, parental behaviour, foraging and many other aspects of adult behaviour to differences in ecology. This brought coherence to a field that had provided a collection of attractive cases for television programmes but did not otherwise seem related to each other. Comparable coherence can be brought to the great variation in the ways in which adult behaviour can develop.

Systems of behaviour that serve different biological functions would not be expected to develop in the same way. In particular, the role of experience is likely to vary considerably from one behavioural system to another. In predatory species capturing fast-moving prey requires considerable learning and practice to be successful. The osprey snatching trout from water does not develop that ability overnight. Animals that rely upon highly sophisticated predatory skills, such as birds of prey, suffer high mortality when young as a result of their incompetence and those that survive are often unable to breed for years; this is because they have to acquire and hone their skills before they can capture enough prey to feed offspring in addition to themselves. In such cases, a combination of different developmental processes is required in order to generate the highly tuned skills seen in the adult.

The developmental processes that make learning, like behavioural imprinting, easier at the beginning of a sensitive period are timed to correspond with changes in the ecology of the developing individual. The processes that bring the sensitive period to an end are often related to the gathering of crucial information, such as the physical appearance of the individual's mother or close kin. Consequently, these processes normally do not terminate the sensitive period until that information has been gathered. In the unpredictable real world, the age when the individual can acquire crucial knowledge is variable; the design of the developmental process reflects that uncertainty.

In contrast to those processes fine-tuned by experience, cleaning the body is not generally something that requires special skills tailored to local conditions. Indeed, grooming by mammals has almost all the various defining characteristics of the old-fashioned notion of instinct. In rodents the duration of the elliptical stroke with the two forepaws which the rodent uses to clean its face is proportional to the size of the species; the bigger the species the longer the stroking movement takes. This is not simply a matter of physics. The bigger-bodied species are not slower in their grooming movements simply because their limbs are heavier; a baby rat grooms at exactly the same rate as an adult rat even though it is a tenth of the size (Berridge, 1994). Moreover, young rodents perform these grooming movements at an age when their mother normally cleans them and before their behaviour patterns are needed for cleaning their own bodies. Rodent grooming is, in other words, a species-typical, stereotyped system of behaviour that develops before it is of any use to the individual.

Midway between the extreme cases are those in which any individual is capable of behaving in a variety of ways, but the way in which it actually behaves is triggered by a cue which it received during a sensitive period early in development. A well-known example is provided by the social insects which may adopt a variety of different forms and behaviour depending on their nutrition received early in life. In such cases the cue may trigger the expression of a set of genes as has been clearly demonstrated in the honeybee (Evans and Wheeler, 2000). In this way complex programmes of development are launched and the one who is the reproductive queen looks and behaves quite differently from her sterile sisters. Such examples emphasise how necessary it is to wean ourselves away from the confused and utterly false idea that genes give rise to instincts and experience gives rise to acquired behaviour.

These conclusions about the muddled use of instinct (and with it, innate and inborn) do not mean that the expression of behavioural characteristics is, what Salman Rushdie called in another context, a P2C2E – a process too complicated to explain. Nor do they mean that such expression cannot be subject to Darwinian evolution when critical environmental conditions are stable from one generation to the next. Nor do they mean that phenotypic expression is totally dependent on the environment. What they do tell us is that if we want to understand developmental processes then we must study them, not merely their antecedents or their consequences. We are dealing with systems. Such systems may be run by simple rules and, if so, are therefore tractable to analysis (Bateson, P. and Horn, 1994). But, as we know from many human games, like chess, the rules may be simple; but the outcomes can be very complex.

Bateson had been castigated for his slowness in understanding the chromosome theory, which so neatly explained segregation and linkage, and for even partially retracting his conversion in the last year of his life (Crowther, 1952). For this he was "relegated to the back lots of scientific history" (Henig, 2000). However, he was interested in the characters of whole organisms. He was aware of the interaction between genes. His sense that the degree of organisation required for the development of an adult organism would not be represented by single particles has a remarkably modern, post-genomic feel to it. The criticism of him for failing to foresee the future of twentieth century molecular biology (in contrast to Archibald Garrod) results from a confusion that has run through these discussions for the past 100 years. We should distinguish in our minds on the one hand the structures, required for the transmission from one generation to the next through the gametes, and on the other hand the developmental systems that lead to the expression of the characters of the whole organism.

Relatively early in his professional career Bateson had an inchoate notion of what he called a "vibratory theory" of development. In an excited letter to his sister Anna, dated 14 September 1891, he wrote: "Divisions between segments, petals etc. are *internodal* lines like those in sand figures made by sound, i.e., lines of maximum vibratory strain, while the mid-segmental lines and the petals, etc. are the *nodal* lines, or places of minimum movement. Hence all the *patterns* and *recurrence of patterns* in animals and plants – hence the perfection of symmetry – hence bilaterally symetrical variation, and the *completeness* of repetition, whether of a part repeated in a radial or linear series etc, etc." (Bateson, B., 1928, p. 43). This idea stayed with him and he wrote about it more extensively in *Problems of Genetics* (1913) and was clearly thinking about these issues into the last year of his life. His would probably be called a systems approach now and the relevance of such thinking to modern biology seems ever more obvious (Bock and Goode, 1998).

In his last published paper, he wrote that while the conception of linkage provided by the chromosome theory probably contains an essential truth, it is in some important respect imperfect (Bateson, W., 1926). He goes on: "What we know of the transmission of family likenesses both in physical and mental attributes is not easily consistent with the theory of random assortment in chromosome groups". A child's characteristics are not a simple blend of its parents' characteristics. Most parents will find some particular likeness between themselves and their child. A daughter might have her mother's hair and her father's shyness, for instance. The child may also have characteristics found in neither parent: a son might have the jaw of his grandmother and the moodiness of his cousin. The shuffling of discrete and supposedly inherited characteristics from one generation to the next is a commonplace of conversation. But what about a whole series of characteristics that sometimes create family likenesses? Are we simply captivated by a single shared characteristic or is something more interesting going on? One idea that might explain family likenesses is that development is a dynamic and selective process. We know that genetic dominance depends a great deal on circumstances and a great many cells die in development. So it is possible that, throughout the process of forming the body, the brain and the behaviour produced by the brain, those characteristics that develop are the ones that work best as an integrated whole. If this were the case, even sharing a very few genes that had a powerful controlling effect on development could lead to startling similarities. Whatever the explanation, Bateson's insight that an understanding of somatic development was needed to understand variation between individuals was way ahead of its time and demands admiration – not contempt.

Both Bateson and Garrod felt that their contributions to evolutionary theory had been neglected. Garrod's sense that the biochemical individuality was crucial is now a part of mainstream thinking. Bateson's sense that the interactions involved in development are crucial is, I suspect, about to have its day. It must be remembered that Bateson with his background in natural history and zoology was primarily interested in whole organisms. Garrod with his medical training was a reductionist – although such a term would not have been recognised in his day. By degrees we have acquired the wisdom to see how both these approaches complement each other.

Acknowledgements

I am grateful to Tim Cox, Anthony Edwards, Milo Keynes and Jim Secord for their comments on an earlier version of this manuscript.

References

Bateson, B. (1928) *William Bateson*. Cambridge: Cambridge University Press.
Bateson, P. (1984) Sudden changes in ontogeny and phylogeny. In Greenberg, G. & Tobach, E. (ed.) *Behavioral Evolution and Integrative Levels*, pp. 155-166. Hillsdale, NJ: Erlbaum.
Bateson, P., Alphey, L., Bulfield, G., Fisher, E., Goodfellow, P., Keverne, E.B., McConnell, I., Owen, M., Radd, M., Smith, J. (2001) *The Use of Genetically Modified Animals*. London: The Royal Society.
Bateson, P., Horn, G. (1994) Imprinting and recognition memory: a neural net model. *Animal Behaviour*, **48**, 695-715.
Bateson, P., Martin, P. (1999) *Design for a Life: How Behaviour Develops*. London: Cape.

Bateson, W. (1894) *Materials for the Study of Variation: Treated with especial regard to Discontinuity in the Origin of Species*. London: Macmillan.

Bateson, W. (1902) *Mendel's Principles of Heredity: A Defence*. Cambridge: Cambridge University Press.

Bateson, W. (1926) Segregation. *Journal of Genetics*, **16**, 201-235.

Beadle, G.W. (1945) Biochemical genetics. *Chemical Reviews*, **37**, 15-96.

Bearn, A.G. (1993) *Archibald Garrod and the Individuality of Man*. Oxford: Clarendon Press.

Berridge, K.C. (1994) The development of action patterns. In Hogan, J.A. & Bolhuis, J.J. (ed.) *Causal Mechanisms of Behavioural Development*, pp. 147-180. Cambridge: Cambridge University Press.

Bock, G.R., Goode, J.A.E. (1998) *The Limits of Reductionism in Biology. Novartis Foundation Symposium 213*. Chichester: Wiley.

Bowler, P.J. (1983) *The Eclipse of Darwinism*. Baltimore: Johns Hopkins University Press.

Brooks, W.K. (1883) *The Laws of Heredity: A Study of the Cause of Variation and the Origin of Living Organisms*. Baltimore: Murphy.

Evolution Committee of the Royal Society (1902) *Report 1: William Bateson and Miss E.R. Saunders*. London: Harrison & Sons.

Crowther, J.G. (1952) *British Scientists of the Twentieth Century*. London: Routledge & Kegan Paul.

Darwin, C. (1859) *On the Origin of Species by Means of Natural Selection*. London: Murray.

Darwin, C. (1872) *The Expression of the Emotions in Man and Animals*. London: Murray.

Ekman, P., Friesen, W.V., O'Sullivan M., et al. (1987) Universals and cultural differences in the judgments of facial expressions of emotion. *Journal of Personality and Social Psychology*, **53**, 712-717.

Eldredge, N. (1995) *Reinventing Darwin*. London: Weidenfeld & Nicolson.

Evans, J.D., Wheeler, D.E. (2000) Expression profiles during honeybee caste determination. *Genome Biology*, **2**, 1-6.

Fisher, R.A. (1930) *The Genetical Theory of Natural Selection*. Oxford: Clarendon Press.

Forsdyke, D.R. (2001) *The Origin of Species Revisited*. Kingston & Montreal: McGill-Queen's University Press.

Freedman, D.G. (1964) Smiling in blind infants and the issue of innate vs acquired. *Journal of Child Psychology and Psychiatry*, **5**, 171-184.

Galton, F. (1892) *Hereditary Genius: An Inquiry into Its Laws and Consequences. 2nd edition*. London: Watts.

Galton, F. (1897) The average contribution of each several ancestor to the total heritage of the offspring. *Proceedings of the Royal Society*, **61**, 401-413.

Garrod, A.E. (1902) The incidence of alkaptonuria: a study in chemical individuality. *Lancet*, 13 December, 1616-1620.

Goldschmidt, R. (1940) *The Material Basis of Evolution*. New Haven, CT: Yale University Press.

Gould, S.J. (2002) *The Structure of Evolutionary Theory*. Cambridge, MA: Harvard University Press.

Gwinner, E. (1996) Circadian and circannual programmes in avian migration. *Journal of Experimental Biology*, **199**, 39-48.

Hailman, J.P. (1987) The ontogeny of an instinct. The pecking response in chicks of the laughing gull (*Larus atricilla* L.) and related species. *Behaviour Supplement*, **15**, 1-159.

Harris, H. (1963) *Garrod's Inborn Errors of Metabolism*. London: Oxford University Press.

Henig, R.M. (2000) *A Monk and Two Peas: The Story of Gregor Mendel and the Discovery of Genetics*. London: Weidenfeld & Nicolson.

Jervis, G.A. (1953) Phenylpyruvic oligophrenia: deficiency of phenylalanine oxidising system. *Proceedings of the Society of Experimental Biology*, **82**, 514-520.

Mayr, E. (1942) *Systematics and the Origin of Species*. New York: Columbia University Press.

Olby, R. (1987) William Bateson's introduction of Mendelism to England: a reassessment. *British Journal of the History of Science*, **20**, 399-420.

Pearson, K. (1898) Mathematical contributions to the Theory of evolution. On the Law of Ancestral Heredity. *Proceedings of the Royal Society*, **62**, 386-412.

Pfaff, D.W. (1997) Hormones, genes, and behavior. *Proceedings of the National Academy of Sciences, USA*, **94**, 14213-14216.

Provine, W.B. (1971) *The Origins of Theoretical Population Genetics*. Chicago: University of Chicago Press.

Troster, H., Brambring, M. (1992) Early social-emotional development in blind infants. *Child Care Health and Development*, **18**, 207-227.

Vollrath, F., Downes, M., Krackow, S. (1997) Design variability in web geometry of an orb-weaving spider. *Physiology and Behavior*, **62**, 735-743.

Wright, S. (1930) Evolution in Mendelian populations. *Genetics*, **16**, 97-159.

Yule, G.U. (1902) Mendel's laws and their probable relations to intra-racial heredity. *New Phytologist*, **1**, 193-207, 222-238.

Human Genetics from 1950

6. Linkage and Allelic Association

Newton E. Morton

Abstract

The first century of Mendelism brought success in the analysis of major genes by linkage and allelic association, leading to positional cloning and, ultimately, to the Human Genome Project. In contrast, genes of lesser effect (oligogenes) present problems that have not been solved and are left to the next century. This is partly because larger samples are required, but mostly because of inherent limitations in the nonparametric methods that have been used and failure to adapt parametric methods to oligogenes. The history of accomplishments and the causes of difficulties are reviewed.

At the dawn of its second century genetics accepts gradation between major genes and polygenes and an intimate connection between linkage and allelic association. Half a century ago these propositions were contentious. Fisher (1918) devised a calculus of variances for quantitative traits that united Mendelism with biometry and became a discipline for plant and animal breeding and behaviour genetics. Mather (1949) preferred variances to gene identification and briefly postulated that polygenes lie in heterochromatin while major genes reside in euchromatin (Mather, 1944). Polymorphism was believed to be rare and maintained by selection (Ford, 1940). In conformity to that dictum, population geneticists attributed allelic association to epistatic selection acting in mathematically clever ways that could not be tested (Lewontin and Kojima, 1960; Arunachalam and Owen, 1971).

The DNA revolution overturned this selectionist view by revealing about three million polymorphisms in the human genome, only a small proportion of which can be maintained by selection. There is no objective dividing line in the continuum between polygenes and major genes. Intermediate oligogenes, the "leading factors" of Wright (1968), are indistinguishable from polygenes by crude methods, but not fundamentally different from major genes by more powerful techniques (Figure 6.1). Mutation rates, selection pressures, penetrance, and gene frequencies vary along this continuum, but do not alter questions about location, sequence, and effect. Allelic association is closely related to linkage, with added variation from time, chance and selection. Both allelic association and linkage reflect location in the DNA sequence and derive from this their utility to identify genes that affect a particular phenotype. The connection is close enough to preclude specialisation in either phenomenon, but loose enough to explain why linkage was studied before allelic association, and major genes before oligogenes.

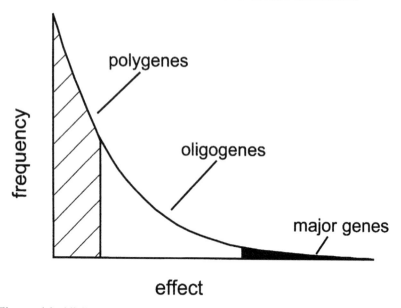

Figure 6.1: Allelic classes: the postulated inverse relation between frequency of contributory alleles and their effect. *Source*: From Wright, S., *Evolution and the Genetics of Populations*. University of Chicago Press, Chicago, 1968.

Linkage of Major Genes

Bernstein (1931) was the first to realise that linkage could be detected in human pedigrees by taking the product of frequencies that must be in coupling or repulsion, whatever the phase of linkage. His method was soon modified by Wiener (1932), Hogben (1934), and Haldane (1934), but was superseded by the elegant maximum likelihood *u* scores of Fisher (1935a) and Finney (1940). These developments were at a time when clinical genetics was in its infancy and few polymorphisms were known that could serve as markers for each other or for major disease loci. The motive for this handful of scholars was intellectual curiosity and familiarity with the power of linkage to solve genetic problems in other organisms. Fisher (1935b) foresaw application of linkage to genetic counselling, but Penrose (1942) expressed a consensus:

> The detection of linkage in man, already advanced as far as the sex chromosome pair is concerned, promises to become a branch of science which, like astronomy, is aesthetically stimulating but must not be expected to have practical uses obvious to the layman.

These pioneers were too modest about the height of their shoulders on which we stand.

Defeat of the Nazis liberated human genetics from its mesalliance with eugenics. Clinical genetics was invigorated, blood grouping flourished, and three autosomal linkages were soon discovered: the Lutheran blood group (LU) and ABH secretion (SE) by Mohr (1951); one form of elliptocytosis (EL1) and the Rhesus blood group (RH) by Lawler (1954); and the nail-patella syndrome (NPS1) with the ABO blood group by Renwick and Lawler (1955). In this short time four of the handful of known human polymorphisms were shown to be linked either to another polymorphism or to a rare disease gene with high penetrance. These antigenic polymorphisms have a phenotype system in which the only complication is recessivity of a blank allele, an ambiguity that is avoided by DNA markers. Genes with high penetrance are called *major*, whether rare or polymorphic. Until the end of the twentieth century they were the only genes that could reliably be used to detect linkage.

Methods for linkage analysis were initially inappropriate even for these simple phenotype systems. They depended on large-sample theory that can be misleading in realistically small samples. They were efficient only in the limit for loose linkage, where power is discouragingly low, and required calculations that were impractical for large pedigrees. Information from sibships with known and unknown linkage phase could not be combined, because the scores were not related to recombination in a simple way. Haldane and Smith (1947) indicated a solution to this problem in their analysis of linkage between haemophilia and colour blindness on the X chromosome, but their introduction of probability ratios was obscured by Bayesian considerations that gave nonadditive scores, required numerical integration, and were of unknown reliability and power. The clue they provided led to sequential analysis, a technique developed in the 1940s to optimise the statistical procedure by which a consignment of bombs was selected or rejected by test explosion of a random sample. If the sample were too large, too few bombs would be left for the war. If the sample were too small, men and machines would be wasted on duds. Sequential analysis introduced two probabilities specifying acceptable and unacceptable parameters that are discriminated with specified type I and type II errors by the smallest mean number of observations. Military secrecy delayed publication of this work (Wald, 1947), which by coincidence was released from censorship the same year that Haldane and Smith first applied lods to analysis of linkage. For simplicity, sequential analysis was developed in terms of a preassigned parameter, say θ, that could be the recombination frequency. This gives tight bounds to power and mean sample size. Estimation by maximising the likelihood with respect to a single parameter complicates these bounds but does not affect the type I error (Collins and Morton, 1991).

The first paper on sequential analysis of linkage tabulated lods for nuclear families with phase known or unknown, derived the prior probability distribution of θ, and concluded from this and the operating characteristics that a lod of 3 is required to assure that at least 95 per cent of significant tests are true (Morton, 1955). In a short time lods were used to disprove earlier claims of autosomal and partial sex linkage based on permissive significance levels (Morton, 1957) and to show that elliptocytosis involves different dominant genes in different pedigrees (Figure 6.2), leading to the perception that "linkage studies have great value in the detection and analysis of genetic heterogeneity, the recognition of which may help to resolve biochemical and clinical heterogeneity" (Morton, 1956). Today, we could add that recognition of genetic heterogeneity is a demonstrated help in diagnosis, both clinical and prenatal, and promises to be invaluable in gene therapy and other forms of prevention and treatment targeted to specific causes.

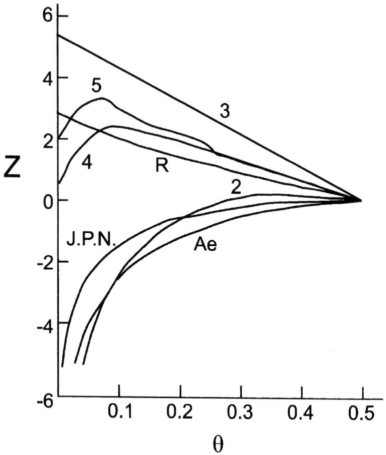

Figure 6.2: Lods for dominant elliptocytosis and RH Pedigrees 3, 4, 5, and R are EL1 (EPB41), closely linked to RH. Pedigrees 2, Ae, and J.P.N. are RH-unlinked.

Early studies presented a standard lod table with values of θ from 0 to 0.5 for a given pair of loci, from which the likelihood could be recovered. Smith (1963) introduced a powerful test of heterogeneity in terms of α, the proportion of families linked to the marker. The importance of sex differences in recombination was recognised (Smith, 1954; Renwick, 1968). Discovery of polymorphisms that could be used as linkage markers progressed from proteins typed by starch gel electrophoresis (Smithies, 1955) to DNA markers (Solomon and Bodmer, 1979; Botstein et al., 1980). The linkage map grew exponentially, and lods for a single pair of loci were replaced by lods on location in a map relative to multiple markers. It became customary to use the equivalences $Z = \chi^2_1/(2 \ln 10)$ and $\chi^2_1 = 2 \ln \{P (\text{data} |\hat{S}) / P (\text{data} |\infty)]$, where Z is a lod expressed conventionally in common logarithms and $\hat{S}$ is the maximum likelihood estimate of location on the linkage map, with $S = \infty$ signifying a location off the map corresponding to $\theta = 0.5$ under H_0. The fundamental theorem for probability ratios, whether in fixed samples or more efficiently in sequential samples, is retained: P $(Z > \log A \mid H_0) \leq 1/A$, with no large-sample approximation. Applications to pedigrees was facilitated by computer programs that analysed the same genetic model with increasing sophistication (Ott, 1974; Lathrop et al., 1985; Lander and Green, 1987; Kruglyak et al., 1996).

Success of major locus linkage revolutionised human genetics. By localising genes to a small interval, typically no more than a few centimorgans (cM), it led to gene identification, cloning, and sequencing. This positional cloning underlies molecular biology and provides clinical genetics with precise tools for diagnosis of cases and carriers. Then, by creating the first connected maps of human chromosomes and demonstrating their utility, major locus linkage led to the Human Genome Project, the goal of which is to sequence not only genes revealed by linkage, but the whole genome. Although, at present, still in draft for most chromosomes, with thousands of gaps and sequence errors, the draft is evolving towards the final map that will guide the next century of Mendelism.

Linkage of Oligogenes

Oligogenes, whether rare or common, have low penetrance and only microphenic effects on liability or a quantitative trait (i.e., less than 1 standard deviation). Identification is not hopeless, but the methods that have been so successful with major genes gave disappointing results until the end of the last century. Human geneticists disagree about whether the problem is with the methods or the way they have been applied. Computer programs for multilocus linkage analysis assume that segregation parameters (frequency of the linked gene, dominance, penetrance in different liability classes, and residual variability due to other loci and family environment) have been determined by analysis that makes appropriate allowance for ascertainment. Given successful analysis, tests on major loci are robust to selection through multiple affected. Although feasible for major genes, prior segregation analysis gives ambiguous results for the small effects of oligogenes. Combined segregation and linkage analysis is more promising, but so far it is limited to single markers and/or random ascertainment (Shields et al., 1994; Guo and Thompson, 1992). Until these limitations are overcome, human genetics is dependent on weakly parametric methods that summarise the genetic model by variance components and identity by descent (IBD) without separating gene frequency and effect and, therefore, with no possibility of modelling ascertainment through affection status. The current literature is full of controversy about whether means, variances, and correlations should be defined on random samples, selected families, or a medley, and the evidence comes from arbitrary simulation, unsupported by genetic probabilities under a realistic ascertainment model. In partial compensation, weakly parametric methods can be applied to extremely nonrandom sampling of affected relatives for which no ascertainment measure has been devised, although there are solutions when all unaffected sibs are included but not necessarily typed (Morton et al., 1991).

Penrose (1935) was the first to recognise that linkage might be detected for a complex trait in pairs of sibs, even if the parents were untested. The promise of parametric lods with major loci made for a cool welcome:

No attempt will be made to treat 'linkage' tests in which the basis of either character is not a single Mendelian factor. If the basis of one or both conditions is multifactoral or unknown, 'linkage' is at best ambiguous and generally cannot be distinguished from any other phenotypic correlation which varies among families. The exploration of these complicated situations may be of some interest, but to include such characters on fancied 'linkage' maps, as some authors have done, is to depreciate the linkage maps that have been determined with some precision in other organisms. (Morton, 1955).

The next generation related the approach of Penrose to identity by descent (Haseman and Elston, 1972) for affected pairs, affection dichotomy, and quantitative traits. Lander and Green (1987) extended this to inheritance vectors for multiple markers, which were generalised from sib pairs to variance components in random pedigrees (Amos, 1994; Blangero and Almasy, 1997). Despite these and other advances, oligogenic linkage has had only modest success in genome scans and candidate regions. Localisation is usually to an interval of at least 20 centimorgans (cM). Although many linkages have ultimately led to positional cloning, others have not been confirmed. The genetic parameters that are known for major genes are unknown for oligogenes, and so there is no prediction of the linkage probability among significant tests, which often have more than the single degree of freedom on which lod theory is based. Evidence may be combined over studies, despite the diversity of markers and phenotypes, but with less conclusive results than for major loci (Lonjou et al., 2001).

The necessity for combination of evidence on studies is apparent for every complex disease. Ornstein-Uhlenbeck theory for Brownian motion predicts the probability of at least one type I error in a genome scan under strong assumptions of a stationary process in an indefinitely large sample, each locus tested independently with 1 degree of freedom, equally spaced loci, and infinite density (Lander and Kruglyak, 1995). All genome scans violate some of their assumptions, which do not predict what proportion of significant results are type I errors.

The dilemma is presented forcefully by psychiatric genetics. About 20 genome scans have been conducted for schizophrenia with stringent phenotypic criteria, but no candidate region has been conclusively demonstrated. On the contrary, a translocation in 1q42 is associated with a range of psychiatric disorders, including schizophrenia, recurrent major depression, and bipolar disorder, and also P300 disturbance in latency and amplitude without psychiatric symptoms (Blackwood et al., 2001). Do the two brain-expressed loci identified by this translocation have oligogenic variation? This uncertainty points to the questionable utility of narrow clinical definition and the necessity to combine evidence over studies. Considering that detection of oligogenes is still so difficult after 65 years of development, confirmation of one or more oligogenes for many complex diseases is a remarkable achievement that will be repeated more often as linkage analysis becomes more rigorous, likelihood-based combination of evidence is perceived to be indispensable, and allelic association is used adaptively to complement linkage.

Table 6.1   Differences between linkage and allelic association.

Attribute	Linkage	Allelic association
Allele specific	0	+
Isolated cases informative	0	+
Case-control informative	0	+
Homozygous parent informative	0	+
Expressed as coupling frequency	0	+
Time dependent	0	+
Drift dependent	0	+
Selection/mutation sensitive	0	+
Linkage phases equiprobable	+	0
Sex specific	+	0
Expressed as IBD	+	0

Allelic Association of Major Genes

Allelic association (often called linkage disequilibrium, or LD) resembles linkage in being dependent on recombination, although influenced by other factors. Both phenomena map to the same locations, and this evidence may be combined for positional cloning. However, there are many differences (Table 6.1). Linkage has exploited the high heterozygosity of microsatellites, whereas allelic association benefits from the greater density of single nucleotide polymorphisms (SNPs). Although several SNPs define haplotypes that give as much heterozygosity as a microsatellite, identification of these haplotypes requires complete typing and is error-prone in the absence of family data or monosomic cell hybrids (Douglas et al., 2001). Because the exponential decrease of allelic association is proportional to time as well as to recombination, the resolution of LD exceeds linkage. In compensation, its power is greatly diminished beyond 1 cM, which in humans is roughly 1 megabase (Mb). The many other differences make LD complementary to linkage, with greatest value after a candidate region has been suggested by linkage, cytogenetics, or functional considerations.

Recent success of positional cloning by LD was made possible by contributions during the last century. Although applications had to await discovery of dense polymorphisms, they depended on theory that began when Robbins (1918) derived from considerations of all genotypes the approach to equilibrium for pairs of neutral diallelic loci in an infinite population, assuming constant allele frequencies and changing haplotype frequencies. Malécot (1948) showed that recurrence for haplotype probabilities could be derived directly. Bennett (1965) and Hill (1974) derived maximum likelihood estimates of haplotype frequencies from samples of diplotypes, and the stage was set to capitalise on LD when enough polymorphisms became available.

Major disease loci are an especially favourable target for LD. They are usually studied in families with multiple cases, and so a haplotype shared by affected relatives may be readily identified. The disease genes are often maintained by recurrent mutations, only a few of which leave many descendants. Affected relatives who do not share a haplotype provide evidence of phenocopies or other loci to be identified. Even if the proportion of unlinked cases is high, LD allows mutation of a specific locus to be localised, cloned, and sequenced by a study of a small number of families showing linkage to the same region. These principles are now familiar to all human geneticists, but their recognition has a short history. Populations that experience a severe reduction in size (bottleneck) because of social or geographic isolation or some natural disaster, war, or foundation by a small number of migrants have reduced genetic complexity and are valuable if there current size provides enough cases (de la Chapelle and Wright, 1998).

Positional cloning of major genes localised by LD began towards the end of the last century when the markers were restriction fragment length polymorphisms (RFLPs), but accelerated when microsatellites and SNPs were identified by the polymerase chain reaction (PCR). There is still innovation with methods to use LD most effectively. Usually, the data consist of families with multiple cases that share a short haplotype, together with a control sample that is also haplotyped. There is no ambiguity about the presence of the major gene in case haplotypes and its absence from controls. Methods based on composite likelihood have been developed that allow for case enrichment in terms of the association probability ρ for a pair of diallelic loci (Collins and Morton, 1998), using evolutionary theory for the number of generations t back to founders in whom a unique mutation was limited to one haplotype. Equilibrium theory is patently inappropriate because of the small value of t, usually less than 100 generations for a deleterious allele. Composite likelihood commonly localises a major gene within 100 kb, an order of magnitude better than linkage. Alternative approaches attempt to model the evolutionary variance from inadequate data. The only successes have been on loci at predetermined positions (Morris et al., 2000). The heavy subjective element in these Bayesian models raises doubt of their untested utility in predicting an unknown location. Multi-marker haplotypes reflect LD, although they are distorted by the same factors (including recombination, mutation, gene conversion and drift) that affect composite likelihood for single markers.

Positional cloning, high-resolution linkage, and evolutionary inference would be facilitated by an LD map in which distances are additive and population differences have been standardised. LD mapping is at the stage of linkage maps nearly a century ago, with the same promise but greater uncertainty about variation within the genome and between populations. Gaps and errors in the human sequence are a great obstacle. Recently, the FRAXE region of Xq28 was displaced 75 Mb in two successive draft sequences. With present sequencing algorithms and efforts it will be years before the promise of sequence-base linkage and LD maps is realised. This makes positional cloning of major loci inefficient, but does not prevent its success. Every month new major genes are identified, even for highly heterogeneous diseases such as deafness and mental retardation that were the despair of geneticists a few years ago.

Allelic Association of Oligogenes

The strategy of identifying a candidate region by linkage and refining it by LD has been brilliantly successful for major genes. In contrast, oligogenes have been much more difficult: as with linkage, there is controversy about whether LD methods or the genetic problem is responsible. The short history of oligogene LD begins with the last decade. By then, the evolutionary theory had been elaborated to include selection, drift and mutation, and many of the infinite number of association measures had been borrowed from statistics. What theory and what metric most facilitate positional cloning? Should the allele frequencies that are now polymorphic be assumed constant in the past, or should their fluctuation and possible fixation be considered? Should equilibrium theory, which is patently inappropriate for disadvantageous genes, be adopted for oligogenes, even if the approach to equilibrium was slow over millenia in which population size and mating patterns were fluctuating in unknown ways?

Finally, should 2-locus haplotypes be dismissed in favour of more complex haplotypes and a theory yet to be developed? Not every human geneticist would answer these questions in the same way. My view is that the approach that is optimal for major loci should be retained for oligogenes so long as the alternatives are poorly defined and unsupported.

A primary difference between major genes and oligogenes is that the former may be unambiguously assigned to a haplotype, whereas oligogenes by definition have effects so small that they cannot be confidently attributed to an individual, let alone to one or the other or both haplotypes. This greatly diminishes the value of haplotyping normal and affected genes unless the oligogene is unambiguously identified by DNA typing, rather than by its phenotypic effect. LD maps for a few short intervals have been constructed, with wide variation in the intensity of LD. There is no international effort to expand this, although it is indispensable for efficient positional cloning. In contrast, a massive effort has begun to define a "haplotype map" for segments of arbitrary content and length from arbitrary populations   (Goldstein, 2001). An LD map gains value if annotated with haplotype frequencies, which contribute to evolutionary studies and in ways yet to be developed may increase resolution of LD mapping. However, a "haplotype map" is an oxymoron unless defined by changes in slope in an LD map. Such changes tend to transform the logarithmic likelihood from its predicted parabola under the Malécot model to a superimposed curve with inflexions more like a step pyramid, with the steps corresponding to recombination events and perhaps to hot spots of recombination. A causal SNP in a relatively flat terrace is distinguishable from neighbouring predictive SNPs by mutation, gene conversion, and recombination that subdivide the deeper clades, if not in an isolate then in an older population with other recombination events and different haplotypes. This is a new world, made possible by genome sequencing and dominated by institution-led research for which the last century did not prepare us. I believe that haplotypes annotated to the LD map will play little role in identifying candidate regions by scanning at low resolution (50 kb), but will become more useful for confirmation at higher resolution where recombination is negligible and so multiple markers are haplotypic. Successful methods to identify causal SNPs will be an extension of major locus approaches:

they will be parametric, multi-marker, nonbayesian, and able to estimate segregation, linkage, and association parameters simultaneously under a correct ascertainment model. It is only fair to add that many of my colleagues think otherwise.

Envoi

The web that connects the handful of pioneers in the last century with the current explosion of genetics can be traced through a few survivors. This is much easier for major genes than for oligogenes and easier for linkage than allelic association. The first pioneers were not human geneticists, since that discipline made a late start compared with experimental genetics. Much of what passed as human genetics under the cloak of eugenics falls between uselessness and maleficence and is embarrassing today. Those of us who lived through that time are convinced that eugenics (like alchemy and astrology) has no legitimate role to play in science or policy (Neel, 2000). Of course, this is not a barrier to personal convictions unless they are unfounded or directive.

The second half of the last century witnessed an increasing proportion of geneticists who chose to leave experimental genetics in favour of their own species. At the same time genetic epidemiology and evolutionary genetics, the major branches of population genetics, began to diverge. Genetic epidemiology is preoccupied with contemporary populations in which some replication is possible. It is family-orientated, disease-biased, and largely inductive. Evolutionary genetics is lineage-orientated, neutrality-biased, and more theoretical. It is concerned with population affinities, allelic genealogies, paleopopulation biology and paleomigration for which general mathematical theories have been developed, but the events themselves were unique and took place in the past under unknown forces of systematic pressure and chance. Divergence tending to inhibit communication was manifest in leading texts, the profession of mathematics or biology, allegiance to different societies, and publication in different journals. Separation was promoted by emphasis on linkage, which is of little interest to evolutionary genetics, but was arrested by allelic association, where the two disciplines cohabit. This century will see hybridisation but not fusion of the two branches.

Genetic epidemiology is faced with increasing emphasis on minor statistical innovations, largely untested on significant genetic problems. Studies outside the laboratory, especially collaborations between different cultures, confront opposition from self-elected group spokesmen and arbiters of political correctness. Distinction between participants in a genetic study and subjects in a clinical trial is essential, but often not made (Morton, 2001). The hope that genetics will be the most significant contributor to human welfare in this century should not blind us to the forces that could reduce genetics to an echo of the triumphs of its first century, primarily on major genes but with a holding attack on polygenes. A science that survived controversy between mendelists and biometricians, distrust of the chromosome theory, Lysenkoism, and eugenics may yet have the vitality to overcome problems of its second century, when genetic research will be extended to populations with different durations and disease exposure, and oligogenes should yield to methods that are now perceived through a glass darkly.

References

Amos C I (1994). Robust variance-components approach for assessing genetic linkage in pedigrees. *Am. J. Hum. Genet.* 54: 535-543.

Arunachalam V and Owen A R G (1971). *Polymorphisms and Linked Loci*. Chapman and Hall , London.

Bennett J H (1965). Estimation of the frequencies of linked gene pairs in random mating populations. *Am. J. Hum. Genet.* 17: 51-53.

Bernstein F (1931). Zur Grundlegung der Vererbung beim Menschen Z *indukt. Abstammungs u Vererbungsl.* 57: 113-138.

Blackwood D H, Fordyce A, Walker M T, St Clair D M, Porteous D J, Muir W J (2001). Schizophrenia and affective disorders – cosegregation with a translocation of chromosome 1q42 that directly disrupts brain-expressed genes: clinical and p300 findings in a family. *Am. J. Hum. Genet.* 69: 428-433.

Blangero J, Almasy L (1997). Multipoint oligogenic analysis of quantitative traits. *Genet. Epidemiol.* 14: 959-964.

Botstein D, White R L, Skolnick M, Davis R W (1980). Construction of a genetic linkage map in man using restriction fragment length polymorphisms. *Am. J. Hum. Genet.* 32: 314-331.

Collins A, Morton N E (1991). Significance of maximal lods. *Ann. Hum. Genet.* 55: 39-41.

Collins A, Morton N E (1998). Mapping a disease locus by allelic association. *Proc. Natl. Acad. Sci. USA.* 95: 1741-1745.

de la Chapelle A, Wright A (1998). Linkage disequilibrium mapping in isolated populations: the example of Finland revisited. *Proc. Natl. Acad. Sci. USA.* 95: 12416-12423.

Douglas J A, Boehnke M, Gillanders E, Trent J M, Gruber S B (2001). Experimentally derived haplotypes substantially increase the efficiency of linkage disequilibrium studies. *Nature Genet.,* 28: 361-364.

Finney D J (1940). The detection of linkage. *Ann. Eugen.* 10: 171-214.

Fisher R A (1918). The correlation between relatives on the supposition of Mendelian inheritance. *Trans. R. Soc. Edinburgh* 52: 399-433.

Fisher R A (1935a). The detection of linkage with "dominant" abnormalities. *Ann. Eugen.* 6: 187-201.

Fisher R A (1935b). Eugenics, academic and practical. *Eugenics Review* 27: 95-100.

Ford E B (1940). Polymorphism and taxonomy. In *The New Systematics.* ed. J. Huxley, pp. 493-513. Clarendon Press, Oxford.

Goldstein D (2001). Islands of linkage disequilibrium. *Nat. Genet.* 29: 109-111.

Guo S W, Thompson E A (1992). A Monte Carlo method for combined segregation and linkage analysis. *Am. J. Hum. Genet.* 51: 1111-1126.

Haldane J B S (1934). Methods for the detection of autosomal linkage in man. *Ann. Eugen.* 6: 26-65.

Haldane J B S, Smith C A B (1947). A new estimate of the linkage between the genes for haemophilia and colour-blindness in man. *Ann. Eugen.* 14: 10-31.

Haseman J K, Elston R C (1972). The investigation of linkage between a quantitative trait and a marker locus. *Behav. Genet.* 2: 3-19.

Hill W G (1974). Estimation of linkage disequilibrium in randomly mating populations. *Heredity.* 33: 229-239.

Hogben L T (1934). The detection of linkage in human families. *Proc. R. Soc. Lond. B.* 114: 340-363.

Kruglyak L, Daly M J, Reeve-Daly M P, Lander E S (1996). Parametric and nonparametric linkage analysis: a unified multipoint approach. *Am. J. Hum. Genet.* 58: 1347-1363.

Lander E, Kruglyak L (1995). Genetic dissection of complex traits: guidelines for interpreting and reporting linkage results. *Nature Genet.* No. 241-247.

Lander E S, Green P (1987). Construction of multilocus genetic maps in humans. *Proc. Natl. Acad. Sci. USA.* 84: 2363-2367.

Lathrop G M, Lalouel J M, Julier C, Ott J (1985). Multilocus linkage analysis in humans: detection of linkage and estimation of recombination. *Am J. Hum. Genet.* 37: 482-498.

Lawler S D (1954). Family studies showing linkage between elliptocytosis and the Rhesus blood group system. Proc. Int. Congr. Genet. IX. Caryologia Suppl., p. 1199.

Lewontin R C, Kojima K (1960). The evolutionary dynamics of complex polymorphisms. *Evolution.* 14: 458-472.

Lonjou C, Barnes K, Chen H, Cookson W O C M, Deichmann K A, Hall I P, Holloway H W, Laitinen T, Palmer L J, Wjst M, Morton N E (2000). A first trial of retrospective collaboration for positional cloning in complex inheritance: assay of the cytokine region on chromosome 5 by the Consortium on Asthma Genetics (COAG). *Proc. Natl. Acad. Sci. USA.* 97: 10942-10947.

Malécot G (1948). *Les mathématiques de l'hérédité.* Masson et Cie, Paris.

Mather K (1944). The genetical activity of heterochromatin. *Proc. R. Soc. B.* 132: 308-332.

Mather K (1949). *Biometrical Genetics.* Methuen, London.

Mohr J (1951). A search for linkage between the Lutheran blood group and other hereditary characters. *Acta Pathol. Microbiol. Scand.* 28: 207-210.

Morris A D, Whittaker J C, Balding D J (2000). Bayesian fine-scale mapping of disease loci by hidden Markov models. *Am. J. Hum. Genet.* 67: 155-169.

Morton N E (1955). Sequential tests for the detection of linkage. *Am. J. Hum. Genet.* 7: 277-318.

Morton N E (1956). The detection and estimation of linkage between the genes for elliptocytosis and the Rh blood type. *Am. J. Hum. Genet.* 8: 80-96.

Morton N E (1957). Further scoring types in sequential linkage tests, with a critical review of autosomal and partial sex linkage in man. *Am. J. Hum. Genet.* 9: 55-75.

Morton N E (2001). Darkness in El Dorado: human genetics on trial. *J. Genet.* 80: 45-52.

Morton N E, Shields D C, Collins A (1991). Genetic epidemiology of complex phenotypes. *Ann. Hum. Genet.* 55: 301-314.

Neel J V (2000). Some ethical issues at the population level raised by "soft" eugenics, euphenics, and isogenics. *Hum. Hered.* 50: 14-21.

Ott J (1974). Estimation of the recombination fraction in human pedigrees – efficient computation of the likelihood for human linkage studies. *Am. J. Hum Genet.* 265: 588-597.

Penrose L S (1935). The detection of autosomal linkage in data which consist of pairs of brothers and sisters of unspecified parentage. *Ann. Eugen.* 6: 133-138.

Penrose L S (1942). Future possibilities in human genetics. *Amer. Naturalist* 76: 165-170.

Renwick J H (1968). Ratio of female to male recombination fraction in man. *Bull. Eur. Soc. Hum. Genet.* 2: 7-12.

Renwick J H, Lawler S D (1955). Genetical linkage between the ABO and nail-patella loci. *Ann. Hum. Genet.* 19: 312-331.

Robbins R B (1918). Some applications of mathematics to breeding problems III. *Genetics.* 3: 375-389.

Shields D C, Ratanachaiyavong S, McGregor A M, Collins A, Morton N E (1994). Combined segregation and linkage analysis of Graves disease with a thyroid autoantibody diathesis. *Am. J. Hum. Genet.* 55: 540-554.

Smith C A B (1954). The separation of the sexes of parents in the detection of human linkage. *Ann. Eugen.* 18: 278-301.

Smith C A B (1963). Testing for heterogeneity of recombination fraction in human genetics. *Ann. Hum. Genet.* 27: 175-182.

Smithies O (1955). Zone electrophoresis in starch gels: group variations in the serum proteins of normal human adults. *Biochem. J.* 61: 629-641.

Solomon E, Bodmer W F (1979). Evolution of sickle variant gene. *Lancet.* ii: 923.

Wald A (1947). *Sequential Analysis.* Wiley, New York.

Wiener A S (1932). Methods of measuring linkage in human genetics, with special reference to blood groups. *Genetics. 17: 335-350* 335-350.

Wright S (1968). *Evolution and the Genetics of Populations.* Vol I. Genetic and Biometric Foundations. The University of Chicago Press.

7. Malaria and Darwinian Selection in Human Populations

Lucio Luzzatto[*]

Summary

The "malaria hypothesis" was stated by J.B.S. Haldane more than half a century ago: long before the Human Genome Project had been dreamed of, much less started. Yet, the fact that many people inherit a relative resistance against the potentially lethal effects of *Plasmodium falciparum* malaria still remains today the best documented example of Darwinian selection shaping the genetic make-up of the human species: specifically, by favouring the increase in population frequency of polymorphic alleles at a number of genetic loci. There are probably several reasons for this. First, with respect to red cell abnormalities, *P. falciparum* is an intracellular parasite; therefore, it is not surprising that almost any significant change in the red cell phenotype might affect the parasite's entry into the red cell, or its growth, or some critical step in its shizogonic cycle. Second, with respect to the immune response, whereas circulating antibodies are not sufficient to protect against *P. falciparum*, acquired immunity is certainly very important, because adults very rarely die of malaria if they have always lived in endemic areas: therefore, any gene that optimises what must be a complex and delicate type of immune response is clearly of great advantage. But the third and possibly the most important reason for the prominence of human genes selected by *P. falciparum* is the very power of malaria selection itself. Indeed, it is obvious that if a condition causes a mortality of, say, 50% of those infected, even a relatively minor protective effect, say 20%, will cause a reduction in mortality sufficient to increase appreciably the frequency of the protective gene within one generation: a protective effect of as little as 2% will cause a reduction in mortality sufficient to increase appreciably the frequency of the protective gene within 10 generations. Moreover, the power of selection is a function not just of mortality in one generation, but also a power function of the number of generations to which it applies. Recent estimates place the spread of malaria in Africa to about 10,000 years ago. Therefore, malaria selection has acted continuously for some 400 generations, thus amplifying enormously its impact: an aspect of this phenomenon of which Haldane had already had the intuition.

In this chapter we will briefly review the evidence for malaria selection of both genes expressed in red cells and genes involved in the immune response. Remarkably, we find examples of almost every possible genetic feature: autosomal genes (e.g., globins) and X-linked genes (G6PD); unique mutation (e.g., in the band 3 gene in South-East Asian ovalocytosis); many allelic mutations indicating convergent evolution (e.g., β-thalassaemia mutations);

---

[*] Work in the author's lab is supported by the Italian Ministry of Health, by the CARIGE Foundation, by AIRC, by the Fondazione San Paolo and by the Ministry of Education

balanced polymorphism through heterozygote advantage with a gene that is semi-lethal in homozygotes (Hb S) versus frequency-dependent selection for mutant homozygotes (Hb C); and point mutations (many examples) versus copy number polymorphism (α, $\alpha\alpha$, $\alpha\alpha\alpha$ at the α globin locus). Different mechanisms of action for these protective genes are emerging.

Introduction

The diseases caused by plasmodial protozoa have been called by a variety of names by the people affected. Interestingly, in African languages the emphasis has been on symptoms: for instance, in Yoruba, *iba*; in Igbo, *iba ocha na-anya* (= fever with yellow eyes). In European languages, instead, the emphasis has been on the presumed cause of the disease, which the French still call *paludisme*. The Romans thought – in a similar vein – that the disease was caused by bad air (*mala aria*): this late Latin word was adopted by the English language and thus became universal.

J.B.S. Haldane[1] first pointed out that an infectious disease that causes a high mortality rate over many generations could be important in shaping human evolution; malaria caused by *P. falciparum* (the other plasmodia are much less lethal) seemed a prime candidate in this respect. Over the past several decades evidence has accumulated that Haldane was right. Indeed, among infectious diseases, *P. falciparum* malaria is one of the best examples of how a parasite can exert a strong selective pressure on its human host. There are at least two strong reasons:

1.   *Plasmodium* is an intracellular parasite of liver cells and of red cells.

2.   The power of selection is great because mortality is high.

Each one of these features is germane, on its own, to natural selection. Taken together, they may explain why *P. falciparum* is a formidable force, which has had a major effect on human evolution in many areas in which it is or has been endemic.

Plasmodium is an Intracellular Parasite of Red Blood Cells

The classical description of an erythrocyte is that of a very altruistic cell that has given up most of its intracellular organelles and metabolic pathways in order to accumulate the highest possible amount of haemoglobin in order to deliver oxygen efficiently to other cells. As a result, there are three main components in the erythrocyte: namely, (a) haemoglobin itself; (b) a limited but essential set of enzymes required for anaerobic glycolysis, redox balance and other necessities; and (c) a functional membrane. It is clear that an intracellular parasite such as *Plasmodium* must interact intimately with each one of these components. Indeed, the intra-erythrocytic schizogonic cycle is characterised by a quite spectacular rate of growth, particularly in the case of *P. falciparum*. Therefore, we can assume that the parasite itself has evolved in a way to optimise its ability to live with, to thrive on and to exploit to its own advantage every one of the features of the host red cell. This must always be true, to some extent, in all cases of intracellular parasitism, but this process of evolutionary adaptation takes on a more compelling quality as a consequence of the minimalist features of the erythrocyte as a host cell that we have already outlined. It is intuitive that the host cell-parasite relationship may be more stringent than if the host cell were, say, a macrophage (as in the case of *Leishmania*). It is also intuitive that the sophisticated complexity of this host cell-parasite relationship must have implications: in short, the closer it is to perfection, the more it may be vulnerable to disturbance or even disruption by small changes. This notion has been put to rigorous testing in human populations, and the remarkable result is that, in each one of what we have called the three components of the red cell, mutations have taken place that have a significant and sometimes major impact on the ability of the parasite to grow.

Several polymorphic mutations in the globin genes are protective against *P. falciparum*

Both structural and quantitative abnormalities in the haemoglobin (Hb) system can affect resistance to *P. falciparum*. Hb S is, of course, the classical example of the former [#3330]; both α- and β-thalassaemia are examples of the latter (some structurally abnormal haemoglobins such as Hb E and Hb Lepore are also protective, probably because they are associated with a β-thalassaemia phenotype). The field studies that have provided the relevant evidence have been reviewed elsewhere.[2-4] Here, we will illustrate this point by a few examples, paying special attention to the likely mechanisms whereby these genes may be resistance factors against *P falciparum*.

Haemoglobin S. A well known paradox in the relationship between haemoglobin S and malaria *in vivo* is that, while heterozygotes very rarely have a life-threatening infection[5,6], homozygotes are at high risk of dying of *P. falciparum* malaria[7]. When parasites are cultured *in vitro* in heterozygous AS** red cells, or even in homozygous SS red cells, their growth is normal, unless the oxygen pressure is reduced, which causes sickling.[8] This clearly indicates that the abnormal haemoglobin in these red cells plays a critical role, but it is not sufficient *per se* to interrupt the parasite cycle. Additional *in vitro* studies have helped to clarify the mechanism of protection. When parasitised Hb AS cells are exposed to autologous monocytes, they are phagocytosed much more efficiently than Hb AA red cells[9]. Moreover, amongst the parasitised red cells it is those that have sickled that are preferentially phagocytosed. Thus, the likely sequence of events is as follows. Parasites in AS cells cause them to sickle[10], not only because they consume oxygen, but also because they decrease the intracellular pH; the sickled AS cells are easy prey to macrophages in the peripheral blood and probably even more within the spleen. Thus, from the point of view of the parasite, the infection of AS cells turns out to be truly suicidal.

Haemoglobin C. In contrast to the geographic distribution of Hb S, which is spread throughout Africa, the Middle East and India, the distribution of Hb C is much more limited, as though it had diffused from an initial ancestor in an area

---

** Hb SS represents the homozygous sickle haemoglobin geotype; AS the simple heterozygous state with one wild-type haemoglobin A allele. Hb CC represents the homozygous state for haemoglobin C, another variant of haemoglobin A caused by a mutation of the glutamic acid residue at position 6 of the β-chain to a lysine; this residue is mutated in sickle cell haemoglobin to valine.

of West Africa, which is today Burkina Faso[11]. In fact, Hb C has been long regarded as an example of a gene that might have become polymorphic in the population of a part of West Africa by genetic drift alone. It has been difficult to prove that malaria selection has operated on this gene, because AC heterozygotes, unlike AS heterozygotes, are not significantly protected. Very recently, however, Modiano et al. have made the remarkable discovery[12] that CC homozygotes are instead very rarely affected by severe, life-threatening malaria. Thus, while the Hb S is a prototype example of balanced polymorphism (since the gene is quasi-lethal in homozygotes), in the case of Hb C homozygotes have only a mild haemolytic disease, which turns out to be an advantage in an environment with heavy malaria transmission. The implications with respect to population dynamics are remarkably different. Because the frequency of homozygotes increases as the square of gene frequency, a significant proportion of the population will be protected from malaria only where the gene frequency is high: a rare example on humans of frequency-dependent selection. The mechanism of protection against *P. falciparum* of CC homozygotes is not yet clear. However, because CC red cells (but not AC red cells) show a tendency to crystallization of haemoglobin, one can speculate that this limits the availability of nutrients for the intracellular parasite. Alternatively, *in vitro* culture studies have suggested that CC red cells may fail to lyse and thus to release merozoites at the appropriate stage of intraerythrocytic development of *P. falciparum* [#5033].

α-thalassaemia. Both the α-globin and the β-globin gene polymorphisms are widespread in malaria-endemic areas. Thus their geographic distributions, although not identical, overlap widely[4]. Both polymorphisms are also characterised by having multiple alleles. However, the polymorphism of the α-globin genes is quite remarkable in another way. Indeed, it is perhaps unique in being a polymorphism in the copy number of a highly expressed, tissue-specific gene. Thus, the number of α-globin genes per chromosome considered "normal" is 2: however, in certain areas of the world, for instance within Nepal and within the South Pacific, there are more people with 1 α-globin gene per chromosome than with 2.

In the Vanuatu archipelago in the South Pacific, which is inhabited by people thought to have migrated originally from South East Asia, there is a gradient in the frequency of the α-thalassaemia gene[13], correlating closely with increasing prevalence of malaria transmission. This is highly suggestive of increasing intensity of Darwinian selection operating on a population with an initial relatively low frequency of this gene.

The Terai region of Nepal, located to the south of the foothills of the Himalayas, has been known to be heavily infested by malaria since remote times. Therefore it has been regarded as virtually uninhabitable by most Nepalese people. As the only exception, the Tharu people have been living in the Terai for centuries, and they were reputed to have an innate resistance to malaria. Since about 1950 the Nepal Malaria Eradication Organization (NMEO) has embarked into a major effort to eradicate malaria, which has achieved a large measure of success. Largely as a result of this, a large and heterogeneous non-Tharu population now inhabits the Terai along with Tharus, creating a rather unique demographic situation. The prevalence of cases of residual malaria is nearly seven times lower among Tharus than in sympatric non-Tharus. This difference applied to both *P. vivax*, which is now much more common, and to *P. falciparum*[14]. Thus, it seems clear that in the Terai holoendemic malaria has caused preferential survival of subjects with α-thal and that this genetic factor has enabled the Tharus as a population to survive for centuries in a malaria-holoendemic area. It can be estimated that the α-thal homozygous state decreases morbidity from malaria by about 10-fold[15]. This is an example of selective evolution towards fixation of an otherwise abnormal gene.

The mechanism of protection is not yet clear in the case of α-thalassaemia. Molecules of parasite origin that land on the red cells surface may play a role[16]; or perhaps altered red-cell membrane band 3 protein may be a target for enhanced antibody binding to α-thalassaemic red cells[17]. On the other hand, it has been suggested that the mechanism of resistance may be, paradoxically, an increase proneness to malaria attacks, possibly favouring the development of immunity. A novel experimental approach, which has now become available for these studies, is the use of mice made transgenic for the relevant human mutant genes[18].

Several polymorphic mutations in one red cell enzyme, those causing G6PD deficiency, are protective against *P. falciparum*

Glucose 6-phosphate dehydrogenase (G6PD) deficiency is the most widely prevalent enzyme disorder of the red cells[19], with population frequencies of 5-20% in many parts of the world and a peak of over 60% in an ancestral population of Kurdish Jews (near fixation or founder effect?[20]). The geographic distribution of the G6PD deficient phenotype is remarkably similar to the epidemiology of malaria as it is now or as it was in the recent past. Since G6PD-deficient subjects are at risk of severe neonatal jaundice and of haemolytic anaemia (#3074), it is difficult to imagine that their genetic trait would reach high frequencies unless it also entails some selective advantage. It has been suggested since 1960 that this might be increased resistance to *P. falciparum* malaria[21,22]. We now know that different polymorphic mutations underlie the G6PD-deficient phenotype in different populations[23]: since these mutations must have arisen independently in genetically disparate people, we can assume that they have been also selected for independently. This further supports the notion that malaria was the selective factor. We have here a good example of evolutionary convergence. Even within a population that is genetically homogeneous, wide variations are observed in the prevalence of G6PD deficiency, depending on the intensity of malaria selection.

Studies in the field have indeed shown that G6PD-deficient children have lower parasite loads[24,25] and a lower incidence of severe malaria[26] than appropriate controls. *In vitro* culture work has demonstrated impaired growth of *P. falciparum* in some studies,[27,28] but not in others[29]. The G6PD gene maps to the X chromosome, and there is some controversy as to whether the relative resistance to this organism is a prerogative of heterozygous females (who are genetic mosaics as a result of X chromosome inactivation) or applies also to hemizygous males[26]. With respect to the mechanism of relative protection against *P. falciparum*, it was shown a long time ago that in infected heterozygous women parasites prefer the G6PD normal red cells[30]. Recently, it has been shown that parasitised G6PD-deficient red cells undergo phagocytosis by macophages at an earlier stage than parasitised normal red cells[29], supporting the notion that also in this case, as for AS heterozygotes, suicidal infection is one protective mechanism.

An interesting point about protective genes expressed in red cells is that haemoglobin has no counterpart in the parasite, whereas in the case of G6PD *P. falciparum* has its own gene encoding this enzyme. In fact, the parasite G6PD gene has highly significant homology to the host cell G6PD gene, but it has also an additional domain[31]. This has been shown to encode 6-phosphogluconate dehydrogenase[32], which is metabolically closely related to G6PD. It is tempting to imagine that this parasite-specific feature, whereby the two enzyme activities are both within the same (bi-functional) protein, might be exploited as a target for anti-malarial agents.

At least one mutation in a red cell membrane protein is protective against *P. falciparum*

Early studies of red cells had investigated a number of surface proteins defined by genetic and serologic analyses as blood groups. The most clear-cut result of these studies was the complete resistance against *P. vivax* of Duffy(-) (Fy/Fy) people.[33] Unfortunately for them, their steadfastness against this species of *Plasmodium* does not extend to *P. falciparum*. The fact that in West Africa the large majority of people have the *Fy/Fy* genotype and *P. vivax* does not exist there must be certainly related to this fact and to each other. Indeed, transmission of *P. vivax* would be obviously interrupted in a geographic area in which there are no susceptible people. It is often assumed that the high prevalence of Duffy(-) people in West Africa is due to selection against Duffy(+) people.[34] However, this would be rather surprising because *P. vivax* infection gives relatively mild disease with little, if any, mortality. An alternative possibility is that the *Fy* gene spread in West Africa through a founder effect, and it was the parasite *P. vivax* that, not finding susceptible hosts, was selected against. Either way, this is a good illustration of the notion, first clearly stated by J.B.S. Haldane (1925), that the evolution of parasites and of their hosts must be regarded always as a co-evolution. In contrast to the case of Duffy, effects on malaria of genetic variation in other blood groups, including ABO, Rh and MN, have been marginal[35].

With respect to *P. falciparum,* by far the most impressive instance of resistance factor in the red cell membrane is that of the the so-called band 3/AE1 anion transporter. This protein is one of the most abundant in the red cell membrane, and a specific mutation (a deletion of codons 400-408, resulting in deletion of 9 amino acids at the boundary between the cytoplasmic and the membrane domain) causes a mild form of haemolytic anaemia associated with a characteristic (oval) red cell morphology, which has a polymorphic frequency in Malaysia and Papua New Guinea,[36] hence the descriptive phrase "South-East Asia ovalocytosis". Clinically, this red cell abnormality protects from cerebral malaria, whereas it may make the anaemia of malaria even worse [#5056; #5053]. Although the mechanism is not completely known, *in vitro* studies suggest that the primary factor may be a decreased rate of invasion of the ovalocytic red cell by the parasite[37].

Genes expressed in cells other than erythrocytes may be also subject to malaria selection

The first port of call of parasites inoculated into a human by the mosquito – the sporozoites – is the liver cell, which has a surface receptor targeted with exquisite specificity by the major surface protein of the sporozoite (CSP)[38]. One could imagine that a genetic variant of the cognate hepatocyte receptor that was unable to bind CSP might be a resistance factor against *P. falciparum*, but this has not yet been explored. From the beginning of malaria infection, another major host response is of course the development of acquired immunity against the parasite, which is highly complex. Indeed, in order for a solid immunity to build up, multiple attacks are required over a period of years[39-41]. In addition, the immunity generated is relative rather than absolute, and it can break down in certain situations (e.g., pregnancy, lack of exposure). Although very numerous antibodies are produced, it appears that cellular immunity may be more important for protection. It is hardly surprising that genetic variation in genes involved in the immune response may affect the development of immunity to malaria. Indeed, at least one HLA allele has been selected by *P. falciparum*[42]. Recent evidence suggests that non-MHC-restricted phospholipid[43] or carbohydrate[44] antigens may be also important in immunity against malaria. The relevant surface molecule is CD1[45], which has little polymorphism. Again, it might be interesting to investigate possible genetic variation of the respective genes in populations living in malaria-hyperendemic areas.

The multiplication of the malaria parasites in the human host during a clinical attack is so massive that disposal of parasites might be just as important as active immunity, and other genes may be involved in this. CD36 is a heavily N-glycosylated trans-membrane protein of 471 aa, which is widely expressed on the surface of epithelial cells, mesenchymal cells and haematopoietic cells. CD36 exhibits genetic polymorphism, and it underlies the Nak[a] blood group on platelets. CD36 tends to localise to caveolae on the cell surface, and it has been shown to bind thrombospondin-1 collagen I, collagen IV long-chain fatty acids, anionic phospholipids and modified LDL. In addition, CD36 binds to certain apoptotic cells and to *P. falciparum*-parasitised red cells. For these reasons, it has been dubbed as a scavenger receptor[46,47]. Indeed, CD36-deficient macrophages show reduced non-opsonic phagocytosis of *P. falciparum* infected red cells[48]. CD36 recognises a *P. falciparum*-encoded ligand with homology to thrombospondin-1. In addition, the adherence of "mature" parisitised red cells to endothelium, characteristic of *P. falciparum* infection, is mediated by CD36. Therefore, it has been suggested that CD36-deficient people may tend to have more severe *P. falciparum* malaria, and this appears to be the case.

The Power of Selection Is Great Because Mortality Is High

The fight against malaria has achieved remarkable successes over the last century. For instance, in the 1930s malaria was fully eradicated in such disparate places as the Southern United States and Central Italy, simply by draining swamps and by the use of larvicides. In the 1950s it was eradicated in the islands of Sardinia, Cyprus and Mauritius by the spraying of insecticides. By similar measures, as well as by the widespread use of chemo-prophylaxis, a considerable measure of control (still short of eradication) was achieved subsequently in countries of Central Asia, as well as in India, Sri Lanka and Venezuela. Nevertheless, such measures have not been applied, or have had hardly any impact, in large parts of the world, particularly in tropical Africa, where the malaria toll in terms of morbidity and mortality is still enormous. Indeed, it is estimated that about 500 million people still live in malaria-endemic areas and that malaria is responsible for 1-3 million deaths/year. Therefore, with respect to malaria-protective genes we are not just speaking of Darwinian selection over an evolutionary time-scale, but of selection still taking place today.

Any selective force that determines a change in the frequency of a gene must do so by decreasing the relative contribution of some or all of the people who have that gene to the next generation. This can, of course, take place in many different ways, but all the mechanisms that can be entertained, including those for which there is experimental support, fall into only two categories: (a) death in the pre-reproductive or in the reproductive age and (b) decreased reproductive activity. It has been suggested that malaria fever can damage sperm cells, or interfere with intercourse;[49] however, it has never been proven that in adults who survive an attack of malaria the rate of reproduction has been significantly reduced (indeed, it would be extremely difficult to obtain such evidence), and in endemic countries today adults rarely die of malaria. On the other hand, it is abundantly clear that in malaria-endemic countries the large majority of deaths from the disease take place in children[50]. Therefore, it is probably safe to assume, at least in first approximation, that the main mechanism of selection by P. falciparum is a high rate of pre-reproductive mortality.

Amongst infectious diseases, malaria is certainly not alone in causing a high rate of mortality: many epidemics have done this in historical times. However, malaria has another important characteristic: in many parts of the world it is not just epidemic, but heavily endemic. The word holoendemic has been coined to indicate those areas where everybody becomes infected. It is estimated that malaria has spread in large tropical and sub-tropical parts of the world for some 10,000 years[51]. It is likely that this was a consequence of the domestication of plants and animals, i.e., the introduction of agriculture. This produced a marked increase in the density of human populations, an essential factor in maintaining malaria transmission. Therefore, unlike with transient epidemics, there has been an opportunity for continuous selection of inherited resistance factors in up to several hundreds human generations.

Concluding Remarks

The malaria parasite has been credited with being a cause of the fall of the Roman Empire. Whereas, clearly, it is impossible to test this notion scientifically, it may not be an exaggeration to say that P. falciparum has played a significant role in shaping human evolution. Both the life cycle and the mechanism whereby the malaria parasite causes death are complex. Therefore, it is not surprising that potential mechanisms of genetic resistance can emerge at the level of several distinct targets within the host, including the erythrocyte and the immune system, as well as local or systemic factors that may influence the development of cerebral malaria, a major cause of mortality. Initially, the information pertaining to these genetic resistance factors was derived from population genetics and from clinical studies. However, it is gratifying that subsequently we have obtained some evidence also on the mechanism whereby these genes operate.

With the evolution of genomics from genetics, new perspectives are opening up. Now that the human genome has been sequenced, we have a complete sequence of the *P. falciparum* genome and a working draft of the *Anopheles* genome. Since the transmission of malaria depends on the interplay among these three genomes, this new knowledge constitutes a powerful tool for progress***. In an interesting recent analysis, the potential practical implications of these advances have been debated. On one hand, it is a sobering consideration that today, in the genomics era, the most effective way to protect ourselves against malaria in endemic areas is still a combination of mosquito nets imbibed in insecticides and chemo-prophylaxis. On the other hand, from the sequence of *P. falciparum* we could identify new immunogens, in the hope of producing a form of vaccination that might improve on natural immunity. Also, by comparing systematically parasite genes *versus* host cell genes we have a powerful new approach to spot potential targets for anti-malarial agents. In addition, we have a frame of reference for the analysis of genetic variation in *P. falciparum*, which will help in tracing the origins and investigating the spread of individual strains, particularly drug-resistant strains.

Genetic host resistance has enabled several populations to survive – albeit at a high cost – the tremendous pressure of malaria selection. It would now be desirable to use the knowledge acquired in this area in order to mimic resistance artificially for the benefit of those who do not have genetic resistance factors. We now have a basis for hoping not only that we can understand in depth the biology of *Plasmodia* and the complex pathogenesis of malaria, but also to catch up with the effective control that has been achieved for other infectious diseases.

References

1. Haldane J.B.S. Disease and evolution. *Ricerca Sci.* 1949; 19, Suppl. I:68-76.
2. Luzzatto L. Genetics of red cells and susceptibility to malaria. *Blood.* 1979; 54:961-976.
3. Miller L.H., Carter R. Innate resistance in malaria. *Exp. Parasitol.* 1976; 40:132-146.
4. Flint J., Harding R.M., Boyce A.J., Clegg J.B. The population genetics of the haemoglobinopathies. *Baillieres Clin. Haematol.* 1998; 11:1-51.
5. Gilles H.M., Fletcher K.A., Hendrickse R.G., Lindner R., Reddu S., Allan N. Glucose-6-phosphate dehydrogenase deficiency, sickling, and malaria in African children in South Western Nigeria. *Lancet.* 1967; 1:138-130.
6. Willcox M., Bjorkman A., Brohult J., Pehrson P.O., Rombo L., Bengtsson E. A. case-control study in northern Liberia of *Plasmodium falciparum* malaria in

*** In fact, genetic resistance factors in the host may be specific for certain genetic variants of the parasite, and there begins to be evidence that this is indeed the case[52] thus, we may have here a good example not just of a parasite influencing evolution of the host, but of the two species actually co-evolving.

haemoglobin S and beta-thalassaemia traits. *Ann. Trop. Med. Parasitol.* 1983; 77:239-246.

7. Adeloye A., Luzzatto L., Edington G.M. Severe malarial infection in a patient with sickle-cell anaemia. *BMJ.* 1972; 2:445-446.

8. Friedman M.J. Erythrocytic mechanism of sickle cell resistance to malaria. *Proc Natl. Acad. Sci. USA.* 1978; 75:1994-1997.

9. Luzzatto L., Pinching A.J. Commentary to R. Nagel - Innate resistance to malaria: the intraerythrocytic cycle. *Blood Cells.* 1990; 16:340-347.

10. Luzzatto L., Nwachuku-Jarrett E.S., Reddy S. Increased sickling of parasitised erythrocytes as mechanism of resistance against malaria in the sickle-cell trait. *Lancet.* 1970; i:319-321.

11. Cavalli-Sforza L.L., Bodmer W.F. *The Genetics of Human Populations.* San Francisco: Freeman; 1971.

12. Modiano D., Luoni G., Sirima B.S., Simpore J., Verra F., Konate A., Rastrelli E., Olivieri A., Calissano C., Paganotti G.M., D'Urbano L., Sanou I., Sawadogo A., Modiano G., Coluzzi M. Haemoglobin C protects against clinical Plasmodium falciparum malaria. *Nature.* 2001; 414:305-308.

13. Flint J., Hill AV., Bowden D.K., Oppenheimer S.J., Sill P.R., Serjeantson S.W., Bana-Koiri J., Bhatia K., Alpers M.P., Boyce A.J., Harding R.M., Clegg J.B. et al. High frequencies of alpha-thalassaemia are the result of natural selection by malaria. *Nature.* 1986; 321:744-750.

14. Terrenato L., Shrestha S., Dixit M., Luzzatto L., Modiano G., Morpurgo G., Arese P. Decreased malaria morbidity in the Tharu people compared to sympatric populations in Nepal. *Ann.Trop. Med. Parasit.* 1988; 82:1-11.

15. Modiano G., Morpurgo G., Terrenato L., Novelletto A., Di Rienzo A., Colombo B., Purpura M., Mariani M., Santachiara-Benerecetti S., Brega A., Dixit K.A., Shrestha S.L., Lania A., Wanachiwanawin W., Luzzatto L. Protection against malaria moribidity: near-fixation of the α-thalassaemia gene in a Nepalese population. *Am. J. Hum. Genet.* 1991; 48:390-397.

16. Luzzi G.A., Merry A.H., Newbold C.I., Marsh K., Pasvol G., Weatherall D.J. Surface antigen expression on *Plasmodium falciparum* infected erythrocytes is modified in α- and β-thalassemia. *J. Exp. Med.* 1991; 173:785-791.

17. Williams T.N., Weatherall D.J., Newbold C.I., Shibata H., Furuumi H., Endo T., Fucharoen S., Hamano S., Acharya G.P., Kawasaki T., Fukumaki Y., Olson J.A., Nagel R.L.. The membrane characteristics of *Plasmodium falciparum*-infected and uninfected heterozygous alpha-thalassaemic erythrocytes. *Br. J. Haematol.* 2002; 118:663-670.

18. Shear H.L., Grinberg L., Gilman J., Fabry M.E., Stamatoyannopoulos G., Goldberg D.E., Nagel R.L. Transgenic mice expressing human fetal globin are protected from malaria by a novel mechanism. *Blood.* 1998; 92:2520-2526.

19. Luzzatto L., Mehta A., Vulliamy T. Glucose-6-phosphate dehydrogenase deficiency. In: Scriver C., Beaudet A., Sly W., Valle D., eds. *The Metabolic and Molecular Bases of Inherited Disease.* Vol. 3 (ed 8). New York: McGraw Hill; 2001:4517-4553.

20. Oppenheim A., Jury C.L., Rund D., Vulliamy T.J., Luzzatto L. G6PD Mediterranean accounts for the high prevalence of G6PD deficiency in Kurdish Jews. *Hum. Genet.* 1993; 91:293-294

21. Allison A.C. Glucose 6-phosphate dehydrogenase deficiency in red blood cells of East Africans. *Nature.* 1960; 186:531-532.

22. Motulsky A.G. Metabolic polymorphisms and the role of infectious diseases in human evolution. *Hum. Biol.* 1960; 32:28-62.

23. Vulliamy T., Luzzatto L., Hirono A., Beutler E. Hematologically important mutations: glucose 6-phosphate dehydrogenase. *Blood Cells, Mol. Dis.* 1997; 23:292-303.

24. Bienzle U., Ayeni O., Lucas A.O., Luzzatto L. Glucose-6-phosphate dehydrogenase deficiency and malaria. Greater resistance of females heterozygous for enzyme deficiency and of males with non-deficient variant. *Lancet.* 1972; i:107-110.

25. Zaccaria A., Barbieri D., Castoldi GL, Ferraresi P, Finelli C, Hossfeld DK, Mitelman F, Rosti G, Testoni N, Van den Berghe H. Normal bone marrow karyotype in paroxysmal nocturnal haemoglobinuria. A cooperative European study. *Cancer. Genet. Cytogenet.* 1983; 9:211-215.

26. Ruwende C., Khoo S.C., Snow R.W., Yates S.N., Kwiatkowski D., Gupta S., Warn P., Allsopp C.E., Gilbert S.C., Peschu N., et al. Natural selection of hemi- and heterozygotes for G6PD deficiency in Africa by resistance to severe malaria. *Nature.* 1995; 376:246-249.

27. Roth J.E., Joulin V., Miwa S., Yoshida A., Akatsuka J., Cohen-Solal M., Rosa R. The use of enzymopathic human red cells in the study of malarial parasite glucose metabolism. *Blood.* 1988; 71:1408-1413.

28. Usanga E.A., Luzzatto L. Adaptation of *Plasmodium falciparum* to glucose 6-phosphate dehydrogenase deficient host red cells by production of parasite-encoded enzyme. *Nature.* 1985; 313:793-795.

29. Cappadoro M., Giribaldi G., O'Brien E., Turrini F., Mannu F., Ulliers D., Simula G., Luzzatto L., Arese P. Early phagocytosis of glucose-6-phosphate dehydrogenase (G6PD)-deficient erythrocytes parasitized by *Plasmodium falciparum* may explain malaria protection in G6PD deficiency. *Blood.* 1998; 92:2527-2534.

30. Luzzatto L., Usanga E.A., Reddy S. Glucose 6-phosphate dehydrogenase deficient red cells: resistance to infection by malarial parasites. *Science.* 1969; 164:839-842.

31. O'Brien E., Kurdi-Haidar B., Wanachiwanawin W., Carvajal J.L., Villiamy T.J., Cappadoro M., Mason P.J., Luzzatto L. Cloning of the glucose 6-phosphate dehydrogenase gene from Plasmodium falciparum. *Mol. Biochem. Parasit.* 1994; 64:313-326.

32. Clarke J.L., Scopes D.A., Sodeinde O., Mason P.J., Olson J.A., Nagel R.L. Glucose-6-phosphate dehydrogenase-6-phosphogluconolactonase. A novel bifunctional enzyme in malaria parasites. *Eur. J. Biochem.* 2001; 268:2013-2019.

33. Miller L.H., Mason S.J., Dvorak J.A., McGinniss M.H., Rothman I.K., Clarke J.L., Scopes D.A., Sodeinde O., Mason P.J., Olson J.A., Nagel R.L. Erythrocyte receptors for (*Plasmodium knowlesi*) malaria: Duffy blood group determinants. *Science.* 1975; 189:561-563.

34. Eaton J.W., Wood P.A. Antimalarial red cells. [Review]. *Prog. Clin. Biol. Res.* 1984; 165:395-412.

35. Menozzi P., Piazza A., Cavalli-Sforza, L.L. *The History and Geography of Human Genes*, 1995.

36. Jarolim P., Palek J., Amato D., Hassan K., Sapak P., Nurse G.T., Rubin H.L., Zhai S., Sahr K.E., Liu S..C, Olson J.A., Nagel R.L. Deletion in erythrocyte band 3 gene in malaria-resistant Southeast Asian ovalocytosis. *Proc. Natl. Acad. Sci. USA.* 1991; 88:11022-11026.

37. Dluzewski A.R., Nash G.B., Wilson R.J., Reardon D.M., Gratzer W.B., Foo L.C., Rekhraj V., Chiang G.L., Mak J.W. Invasion of hereditary ovalocytes by *Plasmodium falciparum* in vitro and its relation to intracellular ATP concentration. *Mol. Biochem. Parasitol.* 1992; 55:1-7.

38. Cerami C., Frevert U., Sinnis P., Takacs B., Clavijo P., Santos M.J., Nussenzweig V., Foo L.C., Rekhraj V., Chiang G.L., Mak J.W. The basolateral domain of the hepatocyte plasma membrane bears receptors for the circumsporozoite protein of *Plasmodium falciparum* sporozoites. *Cell*. 1992; 70:1021-1033.

39. Craig A., Scherf A. Molecules on the surface of the *Plasmodium falciparum* infected erythrocyte and their role in malaria pathogenesis and immune evasion. *Mol Biochem. Parasitol.* 2001; 115:129-143.

40. Phillips R.S. Current status of malaria and potential for control. *Clin Microbiol Rev.* 2001; 14:208-226.

41. Mazier D., Nitcheu J., Idrissa-Boubou M. Cerebral malaria and immunogenetics. *Parasite Immunol.* 2000; 22:613-623.

42. Hill A.V.S. The immunogenetics of human infectious diseases. *Ann. Rev. Immunol.* 1998; 16:593-617.

43. Schofield L., Hewitt M.C., Evans K., Siomos M.A., Seeberger P.H. Synthetic GPI as a candidate anti-toxic vaccine in a model of malaria. *Nature*. 2002; 418:785-789.

44. Schofield L., McConville M.J., Hansen D., Campbell A.S., Fraser-Reid B., Grusby M.J., Tachado S.D. CD1d-restricted immunoglobulin G formation to GPI-anchored antigens mediated by NKT cells. *Science*. 1999; 283:225-229.

45. Joyce S., Woods A.S., Yewdell J.W., Bennink J.R., De Silva A.D., Boesteanu A., Balk S.P., Cotter R.J., Brutkiewicz R.R. Natural ligand of mouse CD1d1: cellular glycosylphosphatidylinositol. *Science*. 1998; 279:1541-1544.

46. Boullier A., Bird D.A., Chang M.K., Dennis E.A., Friedman P., Gillotre-Taylor K., Horkko S., Palinski W., Quehenberger O., Shaw P., Steinberg D., Terpstra V., Witztum J.L., Mazier D., Nitcheu J., Idrissa-Boubou M. Scavenger receptors, oxidized LDL, and atherosclerosis. *Ann. NY Acad. Sci.* 2001; 947:214-23.

47. Febbraio M., Hajjar D.P., Silverstein R.L. CD36: a class B scavenger receptor involved in angiogenesis, atherosclerosis, inflammation, and lipid metabolism. J. *Clin. Invest.* 2001; 108:785-791.

48. Pain A., Urban B.C., Kai O., Casals-Pascual C., Shafi J., Marsh K., Roberts D.J. A non-sense mutation in Cd36 gene is associated with protection from severe malaria. *Lancet*. 2001; 357:1502-1503.

49. Eaton J.W., Mucha J.I. Increased fertility in males with the sickle cell trait? *Nature*. 1971; 231:456-457.

50. Bradley D.J. The last and the next hundred years of malariology. *Parassitologia*. 1999; 41:11-18.

51. Mu J., Duan J., Makova K.D., Joy D.A., Huynh C.Q., Branch O.H., Li W.H., Su X.Z. Chromosome-wide SNPs reveal an ancient origin for Plasmodium falciparum. *Nature*. 2002; 418:323-326.

52. Ntoumi F., Rogier C., Dieye A., Trape J.F., Millet P., Mercereau-Puijalon O. Imbalanced distribution of *Plasmodium falciparum* MSP-1 genotypes related to sickle-cell trait. *Mol. Med.* 1997; 3:581-592.

8. Chromosomal Genetics and Evolution

Malcolm A. Ferguson-Smith

The study of chromosomes is central to the discipline of genetics as it provides the physical basis for Mendel's laws of inheritance. Since the early years of the 20th century it has been the student's surest route to an understanding of genetics and biology. The history of cytogenetics has been punctuated by major technical advances occurring regularly every ten years or so. All the important principles of genetics, including segregation of alleles, linkage, recombination, and transcription, can now be observed under the microscope. It is an added bonus that art and science meet in the study of cytogenetics, providing the investigator with the pleasure of attractive and rewarding images.

In this chapter it seems appropriate to review some of the milestones which have led to the flourishing field of molecular cytogenetics. These have contributed in no small measure to the development of the human genome project, to the emerging discipline of functional genomics, and to a better understanding of karyotype evolution and speciation.

Early History of Human Cytogenetics

Cytogenetics made little impact on human genetics before the 1950s. Human chromosomes were probably first seen in sections of cancer tissues by Arnold (1879) and Hansemann (1891). Flemming (1898) estimated that there were about 24 chromosomes in serial sections of normal corneal cells. Quite different results were reported first by de Winiwarter (1912). He found a count of 47 in adult testis material and 48 in fetal ovarian tissue; he concluded that humans had an X/XX sex determining mechanism and this was accepted by numerous authors for at least 20 years. This is despite the work of Painter (1921), who discovered the Y chromosome in human primary spermatocytes and concluded that 48 was the chromosome number in both sexes. It is of interest that he records that he could count only 46 chromosomes in the clearest mitotic figures.

Of course, the chromosomes of other species were also being studied, and Sutton (1903) and Boveri (1903) are credited with being among the first to appreciate that the behaviour of chromosomes mirrored Mendel's laws of inheritance. But it was the work of Morgan, Sturtevant and Bridges (see Bridges, 1916) that established the chromosomal theory of heredity and later produced the first detailed genetic maps in *Drosophila* (Bridges, 1938). The chromosomes of mammalian species also received attention, notably by Koller and Darlington (1934) and Koller (1937) who distinguished "pairing" and "non-pairing" segments of the X and Y chromosomes and who speculated about the possibility of partial sex linkage in humans.

The Emergence of Modern Human Cytogenetics

There matters stood in human cytogenetics until 1956, when Tjio and Levan found consistent counts of 46 in human fetal tissue cultures. This surprising

91

result was due to several technical improvements coming together to address an old question. The first was the rapid development of tissue culture methods in the late 1940s, the second was the use of colchicine by Levan to arrest and accumulate mitoses, and the third was the accidental discovery that, if one mistakenly immersed mitotic cells in water instead of isotonic solution before fixation, the chromosomes could be separated from one another by gentle squashing under a coverslip. It seems that this happy accident with water occurred independently in 1952 to two cytologists (T.C. Hsu and S. Makino). The human diploid number of 46 was confirmed quickly by Ford and Hamerton (1956) in testicular biopsy material, again exploiting the osmotic effects of hypotonic solution before fixation.

The next question to be answered was the sex chromosomal constitution of patients with the Turner and Klinefelter syndromes. Sex chromatin studies had suggested that they were the result of sex reversal in XY and XX individuals respectively. Indeed, results of chromosome analysis in bone marrow of one Klinefelter patient seemed to confirm this (Ford et al., 1958), although later experience now indicates that this individual must have been the first XX male. Such attempts were overshadowed by the dramatic discovery by Lejeune & colleagues (1959) of an extra small chromosome in several cases of Down syndrome. There quickly followed the identification of the usual chromosome complement in Turner syndrome, i.e. 45, X (Ford et al., 1959) and in Klinefelter syndrome, i.e. 47, XXY (Jacobs and Strong, 1959).

These studies were made on either fibroblast cultures or bone marrow samples treated with colchicine. Many were astonished that such gross genetic aberrations could result in viable offspring, albeit variously handicapped. This prompted a widespread search for a chromosomal basis of other seriously disabling conditions. Trisomy 21 Down syndrome was quickly followed by the discovery of trisomy 18 (Edwards et al., 1960) and trisomy 13 (Patau et al., 1960), and, thereafter, various unbalanced chromosomal rearrangements, including the Philadelphia chromosome in chronic myeloid leukaemia (Nowell and Hungerford, 1960). This intense cytogenetic activity was made easier by another important technical advance. Moorehead et al. (1960) discovered that phytohaemagglutinin when added to peripheral blood samples could induce T lymphocytes to divide during 72 hours incubation. With the use of colchicine in the last 2 hours to accumulate mitoses, followed by a short hypotonic treatment before fixation, it was possible to obtain a cell suspension which could be dropped into microscopic slides and allowed to air dry, producing metaphases which were spread out in one optical plane and thus easily analysed (Figure 8.1; see color photo insert following page 118). The procedure was simple, produced results far superior to anything that had been achieved previously and was available to every pathology laboratory. Its introduction was undoubtedly responsible for the rapid development of human cytogenetics in the 1960s. Chromosome preparations are made by essentially the same method even today.

While the simple "solid" staining techniques used in the 1960s were sufficient to detect numerical and gross structural abnormalities, only chromosomes 1, 2, 3, 16 and the Y could be confidently distinguished from the remainder. However, these years saw a number of attempts to improve resolution using careful measurement of chromosome length and centromere position, and the characterisation of variable heteromorphisms involving the centromeric regions of chromosomes 1, 9 and 16; the short arms and satellites of chromosomes 13, 14, 15, 21 and 22; and the long arm of the Y chromosome. The patterns of DNA replication were found to be a characteristic of each chromosome and useful for distiguishing chromosomes of similar size.

The next milestone in human cytogenetics can be traced to the development in 1970 of two new techniques, namely *in situ* hybridisation and chromosome banding through quinacrine fluorescence. Each contributed to the unequivocal identification of every chromosome by what is now known as Giemsa- or G-banding. Pardue and Gall (1970) showed that repetitive DNA could be annealed *in situ* to complementary DNA in standard air-dried chromosome preparations. The denaturation of the chromosomes by alkali and heat required for the annealing process was noted to produce dark and light stained chromosomal regions with Giemsa stain. Various modifications of these treatments by others yielded a series of bands specific for each chromosome (Figure 8.2). Approximately 500 bands could be consistently recognised in each metaphase, and up to 1000 bands could be recognised when prometaphase chromosomes were selected. Independently, Caspersson et al. (1970) found that metaphases stained with quinacrine compounds which intercalate in DNA yielded bright fluorescent bands along the chromosome when viewed by UV light microscopy. The banding pattern proved to be virtually the same as G-banding and equally useful for chromosome identification. Both methods could detect much smaller structural chromosome aberrations than previously. This proved to be a key advantage in the localisation of genes to human chromosomes using interspecific somatic cell hybrids and deletion mapping, which were being used to construct the human gene map in the 1970s. It also led to a major improvement in clinical cytogenetics with the ability to diagnose many more chromosomal syndromes and to advance the field of leukaemia and cancer cytogenetics. It can be noted that prenatal diagnosis of fetal chromosome abnormalities emerged as an important adjunct to genetic counselling during this period.

It is not possible to give a full account of the early history of human cytogenetics here, and the interested reader may find more detailed descriptions of these events in various reviews and texts (e.g., Ferguson-Smith 1991, 1993; Connor and Ferguson-Smith, 1997; Ferguson-Smith and Smith, 2001).

Molecular Cytogenetics

The hybridisation *in situ* of repetitive DNA to chromosomes by Pardue and Gall (1970) heralded the emergence of molecular cytogenetics. Its first use in human cytogenetics was in the localisation of moderately repetitive ribosomal DNA sequences to the nucleolar organiser regions in the satellite stalks of the five human acrocentric chromosomes (Henderson et al., 1972). The techniques at the time used radioisotopes (such as tritium or ^{125}iodine) to label the DNA sequences and depended on their detection by photographic emulsion applied to the microscope slides. The scatter of radioactive disintegrations around regions of hybridisation reduced the precision of the method, and the long exposures required meant that it took several weeks before the success of the experiment could be determined. This, coupled with the need to count silver grains in many metaphases to achieve statistically significant counts above background levels, led to the search for non-isotopic methods of labelling. Nonetheless, by 1981 the advent of molecular cloning of DNA sequences allowed the first single gene loci to be mapped by isotopic method (Harper et al., 1981; Malcolm et al., 1981). In the same year Langer et al. (1981) used biotin modification of DNA probes, detected by a method depending on horseradish peroxidase coupled to avidin. Later, digoxigenin and other haptens were added to DNA probes, and fluorochromes such as fluorescein-isothiocyanate (FITC) and Texas Red coupled to avidin or antidigoxigenin antibody were used for detection of the annealed DNA probes by fluorescence microscopy. More recently, direct labelling using fluorescent dyes coupled to nucleotides have greatly simplified the procedure. Thus was fluorescence *in situ* hybridisation (FISH) added to the cytogeneticist's armamentarium. Its advantages were rapid results, improved resolution, and a range of different colours allowing several DNA probes to be used simultaneously.

In situ hybridisation is based on the principle that when double-stranded DNA is heated it denatures into single-stranded DNA. On cooling, the single-stranded DNA reanneals with its complementary sequence into double-stranded DNA. If a quantity of labelled DNA sequence is denatured and added to denatured chromosomes during the reannealing process, some of the labelled DNA will hybridise to its complementary sequence in the chromosome. Detection of the labelled DNA will identify the chromosomal site of the DNA sequence, i.e., its position on the chromosomal map. FISH can therefore be used to map genes and other genetic markers, and this facility has played a vital part in the Human Genome Project by helping to order cloned sequences of DNA – a procedure known as "checking the tiling path" of the genome sequence. Since 1985, FISH has become the method of choice for assigning a cloned DNA sequence to its position on the chromosomal map. It has also played an important role in diagnostic cytogenetics, particularly for the detection of structural chromosome abnormalities beyond the resolution of conventional G-banding techniques.

The Development of FISH Probes and Their Application to Human Cytogenetics

FISH demands that DNA probe molecules be available in sufficient excess to provide visible signal under the microscope. This is achieved either by cloning the relevant DNA sequence in appropriate plasmid, cosmid or bacterial artificial chromosome (BAC) vectors, or by amplifying the DNA *in vitro* using the polymerase chain reaction (PCR). An account of these and other techniques can be found in appropriate texts (see, for example, Ferguson-Smith and Smith, 2001). It suffices to mention here the main types of DNA probe used in FISH and some of their applications.

Total genomic probes are prepared from DNA extracted from blood, tissues or cell cultures and labelled appropriately. Chromosomes hybridised with these probes show an evenly distributed signal along their length, referred to as *chromosome painting.* They may be used to identify human chromosome material in interspecific-somatic cell hybrids, including radiation reduced cell hybrids.

Chromosome-specific painting probes are prepared from flow sorted chromosomes (Carter et al., 1992) or microdissected chromosomes (Meltzer et al., 1992). Flow sorting involves the separation of individual chromosomes from a fluid suspension of chromosomes stained with two fluorescent dyes which have an affinity for AT rich and GC rich chromosomes respectively. The chromosomes are sorted by a fluorescence-activated cell sorter into groups depending on their size and base pair ratio. Chromosomes from each group are collected in tubes and 300-500 chromosomes are sufficient to produce chromosome-specific DNA amplified and labelled by random-primed PCR (Telenius et al., 1992).

Figure 8.3a shows a flow karyotype in which the chromosomes have been separated in the above manner from a patient carrying an apparently balanced *de novo* reciprocal translocation between chromosomes 2 and 12. When the patient's chromosomes were analysed using chromosome-specific paints for chromosome 2 (green) and chromosome 12 (red), it became clear that the breakpoints of the translocation occurred in the long arms of each chromosome (Figure 8.3b). As the patient exhibited unexplained serious physical and mental handicap, a further study was undertaken using chromosome paints, made from each product of the translocation by chromosome sorting. Figure 8.3c shows the result obtained when the abnormal chromosome 2 (red) and the abnormal chromosome 12 (green) are hybridised to a *normal* metaphase. This reveals a region of chromosome 12 that is not painted by either of the two translocation chromosomes. It may reasonably be concluded that the patient's handicap is due to loss of that part of chromosome 12 during the formation of the translocation.

Multicolour-FISH probes are now available which identify each individual chromosome in a different colour (Schröck et al., 1996) as in Figure 8.4. This is achieved by using 5 fluorochromes in different combinations for each of the 22 types of autosomes and the X and Y sex chromosomes. Digital fluorescence microscopy using a sensitive (CCD) camera and image analysis is required to measure the contribution of each fluorochrome within the chromosome-specific signal detected along each chromosome. A computer classifier uses this information to identify all the chromosomes in a metaphase and any *inter*-chromosomal rearrangement that may be present, such as a translocation. The limitation of M-FISH is that it cannot identify *intra*-chromosomal rearrangements such as inversions. Various techniques have been developed to deal with this limitation. One method, termed colour banding, is illustrated in Figure 8.5, which shows a complex inversion of chromosome 7. The technique uses gibbon painting probes which exploit evolutionary rearrangements that have occurred during the divergence of humans and lesser apes. Another method uses microdissected probes from several regions of the same chromosome to achieve a detailed colour banding pattern along each chromosome.

Chromosome-specific centromeric probes are prepared from alphoid-repetitive DNA located close to the chromosome centromere. Sequences specific for almost all chromosomes can be cloned and amplified in suitable vectors. Chromosomes 13 & 21 and 14 & 22 are the exception as each pair shares identical sequences and so, for detailed analysis of these chromosomes, sequences are cloned from other regions. Centromeric probes have found application in determining chromosome copy number in interphase nuclei. More than 80% of diploid nuclei will show two signals for each chromosome. This can provide a rapid diagnosis for trisomy (as in Down syndrome) or monosomy (as in Turner syndrome), as the hybridisation can be made on uncultured cells taken directly from the individual. The method is used to screen uncultured amniotic fluid cells for fetal trisomy 21 (Figure 8.6) and for similar studies for pre-implantation diagnosis of early embryos.

Single copy DNA sequence probes are DNA sequences cloned in a suitable vector and usually selected from a genomic library of DNA fragments. Their size depends on the vector used, for example, 30-40 kilobases in a cosmid and 120-130 kilobases in a PAC vector. Such DNA clones are the principal reagents used in the project to sequence the human genome and are readily available. In fact, they can be used to produce a series of reference markers at regular intervals along the chromosome for mapping unknown disease genes by linkage analysis. Figure 8.7 shows a metaphase from a patient with a cryptic translocation between the ends of chromosomes 7 and 21. Two closely-linked cosmid markers on the normal chromosome 21 have been separated by the translocation breakpoint so that one (proximal) cosmid (red) remains on the chromosome 21 derivative, while the chromosome 7 derivative has received the other cosmid (green). The red and green cosmids on the normal chromosome 21 are so closely linked that they together generate a yellow signal. Cosmid clones are widely used for the diagnosis of the relatively common microdeletion syndromes (Figure 8.8).

As most structural inter-chromosomal rearrangements involve the ends of chromosomes, and as G-banding and chromosome specific paint probes cannot readily detect duplications or deletion at less than 2-3 megabases, **telomere-specific DNA probes** have been produced which delineate the ends of all human chromosomes and which can be used to detect cryptic translocations. These probes depend on chromosome-specific sequences located at less than 300 kilobases from the chromosome ends. M-FISH allows many of these to be used simultaneously in one hybridisation, providing another useful diagnostic tool for use in clinical practice.

When two cosmid clones are less than 2-3 megabases apart on a metaphase chromosome (as in Figure 8.7), they cannot be distinguished separately. This is because the chromosome fibre is attached to a protein scaffold at many points along its length in such a way that the fibre loops out from its attachments, producing a "bottlebrush-like" structure. Techniques that remove histones from the chromosomes tend to release the DNA from its scaffold, and this finding has been exploited in the preparation of microscope slides in which extended DNA fibres radiate in a halo around each nucleus. Many are present as loops, but when these are broken the DNA fibre is extended even further. DNA labelled probes can be hybridised directly onto extended DNA fibres for FISH analysis, which has the power to resolve distances less than 5 kilobases. In such preparations, one micron is equivalent to 3 kilobases. Figure 8.9 shows the contiguous array of 3 linked cosmids from the MHC locus, each labelled in a different colour and each consisting of approximately 35 kilobases. The technique allows individual genes to be resolved into their respective introns and exons with the various intervals measured in kilobase units. No form of genetic linkage other than DNA sequencing has higher resolution.

The introduction of FISH has led to another important development with particular application to the genetic analysis of cancer tissue. **Comparative genomic hybridisation** (CHG) is used to identify DNA duplications of over 5-10 megabases and deletions in the order of 10-20 megabases in tumours using reverse chromosome painting (Kallioniemi et al., 1992). FITC-labelled total *tumour* DNA (green) is mixed with Texas red-labelled *normal* reference DNA (red) and hybridised to normal metaphases. The relative amounts of tumour and normal DNA that hybridise to a particular chromosome region depend on the number of copies of DNA sequences complementary to that region in the tumour sample. A duplicated region will be revealed by an increased green/red fluorescence ratio, while a deleted region will appear as a region of decreased green/red fluorescence ratio. Fluorescence ratios are determined by digital fluorescence microscopy, in which the relative amounts of green and red fluorescence are measured along the length of each chromosome. The method does not require the direct analysis of cancer chromosomes, which can be technically difficult in most cancers. It has helped in the identification of chromosome aberrations associated with specific tumours and has also been used in the detection of constitutional aberrations in handicapped patients.

Cross-Species Chromosome Homology

When chromosome-specific paint probes from one species are hybridised to the metaphases of another species, regions of homology can be detected with variable levels of efficiency depending on the closeness of the relationship of the two species. Cross-species painting between humans and the great apes reveals that, with two exceptions, each human paint identifies one whole ape chromosome (Figure 8.10). The exceptions are human chromosome 2, which is represented by two separate ape chromosomes, and the presence of one chromosome in the gorilla, which is composed of two blocks equivalent to parts of human chromosomes 5 and 17. It is apparent that chromosome specific DNA is conserved between apes and humans along the entire length of each chromosome. However, repetitive DNA has not been well conserved during evolution. This is demonstrated by the improved resolution achieved when ape paints are used instead of human paints in human chromosome diagnosis. Background signals due to repeated DNA sequences scattered throughout the karyotype are substantially reduced.

G-banding comparisons between the human and great apes show a number of inversions which are not visible by chromosome painting between humans and great apes. However, using the gibbon paint probe set (as in Figure 8.5) many of these inversions become apparent (Ferguson-Smith et al., 2000). This is because there have been numerous interchromosomal rearrangements during the 15 million years that have elapsed during the divergence of the gibbon and human karyotypes. Human paints reveal over 60 separate homologous segments in the gibbon karyotype. In the dog, human paints recognise 74 homologous segments but in the cat the number is only 34, although both species diverged from humans about 46 million years ago. Clearly, the rate of rearrangement is not simply a matter of time. It is believed that the cat karyotype is closer to the ancestral mammalian karyotype than some 50 or so other species that have been examined so far.

Comparison of the genetic maps of mouse and human show that the mouse karyotype is the most rearranged of all mammals. It is estimated that there are over 180 separate homologous segments between mouse and human. Cross-species painting confirms this and also confirms the close correspondence between the homology map produced by painting and the one evident from genetic mapping (Ferguson-Smith, 1997).

Reciprocal chromosome painting has proved very useful in the study of karyotype evolution and phylogeny. It complements both gene mapping and G-banding in that it demonstrates that genetic linkage groups have been conserved intact in species which diverged many million years ago. Chromosome painting provides one of the simplest methods for determining phylogenetic relationships, as it can distinguish between arrangements of segments which are shared by many distantly related species from arrangements of more recent origin which are shared only by a few closely-related species. As the former are more likely to represent the more ancestral situation it is possible to construct trees of relationships between species. This can be illustrated by considering the results of painting with human chromosome paints across a wide range of animals. For example, human chromosome 17 paint hybridises to only one segment of the karyotype in most mammals (with some exceptions including the dog). Human chromosome paints from chromosomes 14 & 15 are associated in one segment in most mammals with the exception of apes and the dog and the same is more or less true for human chromosomes 3 & 21, 12 & 22, and 16 & 19 (see Table 8.1). These associations can be regarded as ancient arrangements when compared to the situation in humans. On the other hand, human chromosomes 5 & 19 are found in association only in artiodactyls and the dolphin, and human chromosomes 3 & 19 are associated only in carnivores, seals and dolphins. The human X has homology to all eutherian Xs and also the long arm of the marsupial X; this must be the best example of an ancestral arrangement whose origin can be traced to the emergence of mammalian sex chromosomes.

Chromosome paints from a number of species have been hybridised to the three pairs of chromosomes of the Indian muntjac, and in some of these species muntjac paints have been used in reciprocal painting. This is summarised in Figure 8.11, in which an idiogram of the female muntjac shows the segments of homology found in human, sheep, cattle, Chinese muntjac, and brown brocket deer. The idiogram acts as a nomogram to provide, at a glance, the comparative homology between the species, including the ancient associations mentioned in the previous paragraph.

As the Indian muntjac is the only extant mammal with such a small number of chromosomes (Figure 8.12), it was one of the first to be used in phylogenetic studies by chromosome painting (reviewed in Yang et al., 1995). The painting results are consistent with the earlier views, based on G-banding, that the ancestor of all extant deer species had a high chromosome number (probably, 2n = 70) and that the small chromosome numbers found in muntjacs and closely related species have evolved largely by a mechanism of *chromosome fusion* (Yang et al. 1997).

A similar study using chromosome-specific paints from the domestic dog provided evidence that the immediate ancestor of the canid family had a smaller number of chromosomes (perhaps about 34) and that the dog karyotype achieved a chromosome number of 78 largely by a process of *chromosome fission* (Graphodatsky et al., 2001). It is not clear as yet how the additional centromeres were generated to enable chromosome numbers to increase from 34 to 78.

Apart from the mouse, the domestic dog has (when compared to humans) the most extensively rearranged karyotype of all mammalian species studied to date. Dog chromosome-specific paints recognise 90 homologous human segments. This level of rearrangement has been used to demonstrate the comparatively high frequency of intrachromosomal rearrangements, mostly inversions, which have occurred during the divergence of humans and other mammals, in particular the cat (Yang et al., 2000). Multiple inversions are the main rearrangements that are found between human and the great apes and have been, presumably, a factor in speciation.

Chromosome painting has made valuable contributions to the development of genetic maps in unmapped species by allowing direct comparison with the maps of well-mapped species such as human and mouse. This has recently been well demonstrated in the domestic dog, in which the large number of chromosomes (78) and the small size and similarity of the smallest 17 pairs has made chromosome identification extremely difficult. However, by chromosome sorting it has been possible to make chromosome paints from every dog chromosome (Yang et al., 1999) and to use the chromosome-specific DNA to map all the genetic linkage and radiation hybrid groups to their respective chromosomes (Sargan et al., 2000). This has provided a secure foundation for the dog map, to which many specific genes and many hundred genetic markers have now been assigned with confidence. It is now possible to begin the genetic analysis of the 350 or so Mendelian disorders and traits which have been identified in the approximately 400 breeds of domestic dog. In the future it may even become possible to find genetic loci which influence the distinctive behavioural traits in some breeds. Such studies could have important implications for human genetics as canine loci will have their homologues in the human. Likewise, our knowledge of the human genome can be expected to assist in the identification of canine homologues of human genetic disease including some cancers.

There is increasing interest in the study of comparative genetics as a tool for the discovery of sequences that regulate and modify human genes. Apart from transcribed gene sequences, it is self-evident that other DNA sequences which are widely conserved across species are likely to have important functions. Analysis of these conserved sequences in "knock-out" transgenic animals may well lead to an understanding of their function. Chromosome painting will play a role in this endeavour and may direct which chromosomes of certain species should first be subjected to complete DNA sequencing. It is to be expected that advances in molecular cytogenetics in general will continue to contribute to such diverse fields as functional genomics, clinical diagnosis, the study of cancer, and phylogenetic relationships.

Acknowledgements

Much of the recent work illustrated in this chapter has been undertaken by members of the Molecular Cytogenetics team at the Department of Clinical Veterinary Medicine, University of Cambridge. In particular, I am grateful to Drs. Fengtang Yang and Willem Rens for their brilliant FISH work, to Patricia O'Brien for chromosome sorting and to Beiyuan Fu and Bruce Milne for skilled assistance in complex techniques.

Table 8.1: Ancient syntenies revealed by association of chromosome segments homologous to human chromosomes in the genomes of various mammals. Chromosome numbers refer to individual human chromosomes.

Species	Ancient syntenies								
	14/15	3/21	12/22	16/19	4/8	7/16	3/19	1/2	5/19
Great Apes	-	-	-	-	-	-	-	-	-
Lesser Apes	-	-	-	-	-	-	-	-	-
Old World Monkeys	+	-	-	-	-	-	-	-	-
New World Monkeys	±	±	-	-	-	-	-	-	-
Lemurs	+	+	+	-	-	-	-	-	-
Cattle	+	+	+	+	+	+	-	+	+
Sheep	+	+	+	+	+	+	-	+	+
Muntjac	+	+	+	+	-	+	-	+	+
Pig	+	+	+	+	-	-	-	-	+
Horse	+	+	+	+	+	+	-	-	-
Zebra	+	+	+	+	+	+	-	+	-
Dolphin	+	+	+	+	-	-	+	-	+
Seal	+	+	+	+	+	-	+	-	-
Cat	+	+	+	+	+	+	+	+	-
Mink	+	+	+	+	+	+	+	-	-
Dog	-	+	+	-	+	+	+	+	-
Panda	-	+	+	+	+	.	+	+	-
Rabbit	+	+	+	+	+	+	-	-	-
Shrew	+	+	-	+	-	-	-	-	-

References

Arnold, T. (1879) Beobachtungen über Kerutheilungen in den Zellen der Geschwulste. *Virchows Arch.* **78**, 279-301.

Boveri, T. (1903) Uber den Einfluss der Samenzelle auf die Larvencharaktere der Echiniden. *Arch. Ent. Der Organismen*, XVI.

Bridges, C.B. (1916) Nondisjunction as a proof of the chromosome theory of heredity. *Genetics* **1**, 1-52, 107-163.

Bridges, C.B. (1938) A revised map of the salivary gland X chromosome of Drosophila melangaster. *J. Heredity* **29**, 11-13.

Carter, N.P., Ferguson-Smith, M.A., Perryman, M.T., Telenius, H., Pelmear, A.H., Leversha, M.A., Glancy, M.T., Wood, S.L., Cook, K., Dyson, H.M., Ferguson-Smith, M.E., Willatt, L.R. (1992) Reverse chromosome painting: a method for the rapid analysis of aberrant chromosomes in clinical cytogenetics. *J. Med. Genet.* **29**, 299-307.

Caspersson, T., Zech, L., Johansson, C. (1970) Differential banding of alkylating fluorochromes in human chromosomes. *Exp. Cell. Res.* **60**, 315-319.

Connor, J.M., Ferguson-Smith, M.A. (1997) *Essential Medical Genetics*, Fifth Edition, Blackwell Science, Oxford.

de Winiwarter, H. (1912) Etudes sur la spermatogenese humaine. *Arch. Biol.* (Liège) **27**, 91-189.

Divane, A., Carter, N.P., Spathas, D.H., Ferguson-Smith, M.A. (1994) Rapid prenatal diagnosis of aneuploidy from uncultured amniotic fluid cells using five-colour fluorescence in situ hybridisation. *Prenatal Diagn.* **14**, 1061-1069.

Edwards, J.H., Harnden, D.G., Cameron, A.H., Crosse, M.V., Wolff, O.H. (1960) A new trisomic syndrome. *Lancet* (i), 787.

Ferguson-Smith, M.A. (1991) Putting the genetics back into cytogenetics. *Am. J. Hum. Genet.* **48**, 179-182.

Ferguson-Smith, M.A. (1993) From chromosome number to chromosome map: the contribution of human cytogenetics to genome mapping. Chapter 1, In: *Chromosomes Today*, Vol. II, Eds. Sumner, A.T. & Chandley, A.C., Chapman & Hall, London, pp.3-19.

Ferguson-Smith, M.A. (1997) Genetic analysis by chromosome sorting and painting: phylogenetic and diagnostic applications. *Eur. J. Hum. Genet.* **5**, 253-265.

Ferguson-Smith, M.A., O'Brien, P.C.M., Rens, W., Yang, F. (2000) Comparative Chromosome Painting, In: *Chromosomes Today*, Vol. 13, Eds. Olmo, E. & Redi, C.A., Birkhäuser Verlag, Switzerland, pp. 259-265.

Ferguson-Smith, M.A., Smith, K. (2001) Cytogenetic Analysis, Chapter 25, In: *Principles and Practice of Medical Genetics*, Eds. Rimoin, D.L., Connor, J.M. & Pyeritz, R.E., 4th Edition, Churchill-Livingstone, New York.

Ferguson-Smith, M.A., Yang, F., O'Brien, P.C.M. (1998) Comparative mapping using chromosome sorting and painting. *ILAR J.* **39**, 68-76.

Flemming, W. (1898) Ueber die chromosomenzahl beim Menshen. *Anatomischer Anzeiger* **14**, 171-174.

Ford, C.E., Hamerton, J.L. (1956) The chromosomes of man. *Nature* **178**, 1020.

Ford, C.E., Jacobs, P.A., Lajtha, L.G. (1958) Human somatic chromosomes. *Nature* **181**, 1565-1568.

Ford, C.E., Jones, K.W., Polani, P.E., De Almeida, J.C., Briggs, J.H. (1959) A sex-chromosome anomaly in a case of gonadal dysgenesis (Turner's syndrome). *Lancet* (i), 711-713.

Graphodatsky, A.S., Yang, F., O'Brien, P.C.M., Perelman, P., Milne, B.S., Serdukova, N., Kawada, S.I., Ferguson-Smith, M.A. (2001) Phylogenetic implications of the 38 putative ancestral chromosome segments for four canid species. *Cytogenet. Cell Genet.* **92**, 243-247.

Hansemann, D. (1891) Uber Pathologische Mitosen. *Virchows Arch.* **123**, 356.

Harper, M.E., Ullrich, A., Saunders, G.F. (1981) Localisation of the human insulin gene to the distal end of the short arm of chromosome 11. *Proc. Natl. Acad. Sci. USA* **78**, 4458-4460.

Henderson, A.S., Warburton, D., Atwood, K.C. (1972) Location of rDNA in the human chromosome complement. *Proc. Natl. Acad. Sci. USA* **69**, 3394-3398.

Hsu, T.C. (1952) Mammalian chromosomes *in vitro*. I. The karyotype of man. *J. Hered.* **43**, 167-172.

Jacobs, P.A., Strong, J.A. (1959) A case of human intersexuality having a possible XXY sex determining mechanism. *Nature* **183**, 302.

Kallioniemi, A., Kallioniemi, O.P., Sudar, D., Rutovitz, D., Gray, J.W., Waldman, F., Pinkel, D. (1992) Comparative genomic hybridisation for molecular cytogenetic analysis of solid tumours. *Science* **258**, 818-821.

Koller, P.C. (1937) The genetical and mechanical properties of sex chromosomes. III. Man. *Proc. Roy. Soc. Edinb. B* **57**, 194-214.

Koller, P.C., Darlington, C.D. (1934) The genetical and mechanical properties of the sex chromosomes. I. *Rattus norvegicus. J. Genet.* **29**, 159-173.

Langer, P.R., Waldrop, A.A., Ward, D.C. (1981) Enzymatic synthesis of biotin-labelled polynucleotides: novel nucleic acid affinity probes. *Proc. Natl. Acad. Sci. USA* **78**, 6633-6637.

Lejeune, J., Gautier, M., Turpin, R. (1959) Les chromosomes somatique des enfants mongoliens. *C. R.. Acad. Des Sci. Paris* **248**, 1721.

Makino, S., Nishimura, I. (1952) Water pre-treatment squash technique. *Stain Technol.* **27**, 1-7.

Malcolm, S., Barton, D., Bentley, D.L., Ferguson-Smith, M.A., Murphy, C.S., Rabbitts, T.S. (1981) Assignment of a IGKV locus for immunoglobulin light chains to the short arm of chromosome 2 (2cen->p13) by in situ hybridisation using a cRNA probe of H101λXH4A. *Cytogenet. Cell Genet.* **32**, 296.

Mann, S.M., Burkin, D.J., Griffin, D.K., Ferguson-Smith, M.A. (1997) A fast, novel approach for DNA fibre fluorescence in situ hybridisation analysis. *Chrom. Res.* **5**, 145-147.

Meltzer, P.S., Guan, X.Y., Burgess, A., Trent, J.M. (1992) Rapid generation of region specific probes by chromosome microdissection and their application. *Nature Genet.* **1**, 24-28.

Moorehead, P.S., Nowell, P.C., Mellman, W.J., Battips, D.M., Hungerford, D.A. (1960) Chromosome preparations of leukocytes cultured from human peripheral blood. *Exp. Cell Res.* **20**, 613-616.

Nowell, P.C., Hungerford, D.A. (1960) A minute chromosome in human chronic granulocytic leukemia. *Science* **132**, 1497.

Painter, T.S. (1921) The Y chromosome in mammals. *Science* **53**, 503.

Pardue, M.L., Gall, J.G. (1970) Chromosomal localisation of mouse satellite DNA. *Science* **168**, 1356-1358.

Patau, K., Smith, D.W., Therman, E., Inhorn, S.L., Wagner, H.P. (1960) Multiple congenital anomalies caused by an extra autosome. *Lancet* (i), 790.

Sargan, D.R., Yang, F., Square, M., Milne, B.S., O'Brien, P.C.M., Ferguson-Smith, M.A. (2000) Use of flow-sorted canine chromosomes in the assignment of canine linkage, radiation hybrid, and syntenic groups to chromosomes: refinement and verification of the comparative chromosome map for dog and human. *Genomics* **69**, 182-195.

Schröck, E., du Manoir, S., Veldman, T., Schoell, B., Wienberg, J., Ferguson-Smith, M.A., Ning, Y., Ledbetter, D., Bar-Am, I., Soenksen, D., Garini, Y., Ried, T.

(1996) Multicolor spectral karyotyping of human chromosomes. *Science* **273**, 494-497.

Sutton, W.S. (1903) The chromosomes in heredity. *Biol. Bull.* **4**, 231-251.

Telenius, H., Pelmear, A.H., Tunnacliffe, A., Carter, N.P., Behmel, H., Ferguson-Smith, M.A., Nordens, K.J., Old, J.M., Pfragner, R., Ponder, B.A.J. (1992) Cytogenetic analysis by chromosome painting using DOP-PCR amplified flow-sorted chromosomes. *Genes Chrom. Cancer* **4**, 257-263.

Tjio, J.H., Levan, A. (1956) The chromosome number of man. *Hereditas* **42**, 1-6.

Yang, F., Carter, N.P., Shi, L., Ferguson-Smith, M.A. (1995) A comparative study of karyotypes of muntjacs by chromosome painting. *Chromosoma* **103**, 642-652.

Yang, F., Graphodatsky, A.S., O'Brien, P.C.M., Colabella, A., Solanky, N., Squire, M., Sargan, D.R., Ferguson-Smith, M.A. (2000) Reciprocal chromosome painting illuminates the history of genome evolution of the domestic cat, dog and human. *Chrom. Res.* **8**, 393-404.

Yang, F., O'Brien, P.C.M., Milne, B.S., Graphodatsky, A.S., Solanky, N., Trifonov, V., Rens, W., Sargan, D., Ferguson-Smith, M.A. (1999) A complete comparative chromosome map for the dog, red fox and human and its integration with canine genetic maps. *Genomics* **62**, 189-202.

Yang, F., O'Brien, P.C.M., Wienberg, J., Ferguson-Smith, M.A. (1997) A reappraisal of the tandem fusion theory of karyotype evolution in the Indian muntjac using chromosome painting. *Chrom. Res.* **5**, 109-117.

9. Mendelian Disorders in Man: The Development of Human Genetics

Timothy M. Cox

Here we examine the origin of biochemical genetics in the earliest years of the twentieth century by Archibald Garrod (1857-1936) and the general study of Mendelian disorders in man. The brief but fruitful scientific relationship between Garrod and William Bateson (1861-1926) – who translated Mendel's work and championed it in the English scientific world – stemmed from their common fascination with evolutionary biology. The product of this collaboration flourished beyond all expectations, and its scientific applications in medicine have had profound consequences for the concept of disease. Genetic research has risen as a dominant influence on the tenor and organisation of the universities and has had some effect on medical education; it continues to provoke much controversy in discussion about clinical teaching and practice and in the wider political scene. Bateson's and Garrod's discoveries have also had far-reaching effects on the understanding of human population genetics and developmental biology.

Biochemical Genetics of Alkaptonuria

Archibald Garrod in his presentation at the Royal Medical and Chirurgical Society of London first hinted at the inborn component of alkaptonuria almost exactly 100 years ago: he presented his lecture in November 1901 and in it first commented on the high frequency of cousin marriages in the parents of affected individuals. Garrod's experimental research into the consequences of that inborn factor is a model of clinical investigation. Garrod deduced the biochemical defect in alkaptonuria by carefully designed feeding experiments in which he noted that he could increase the excretion of homogentisic acid in the urine by feeding protein-rich foods and aromatic acids related to tyrosine. Garrod concluded that alkaptonuria resulted from a failure to break open the benzene ring in the degradative pathway for tyrosine and phenylalanine.

Inborn Errors of Metabolism

Bateson learnt of Garrod's work in 1901 and rapidly perceived the significance of parental consanguinity in alkaptonuria, reporting his own hypothesis of its transmission as a human autosomal recessive trait to the Royal Society in December of that year. Thereafter, Bateson and Garrod corresponded and became friends. Garrod developed his concept of the inborn error of metabolism which, in alkaptonuria, affected the activity of a liver enzyme ("ferment") involved in the scission of the benzene ring of tyrosine.

This conclusion was elegantly articulated in his 1908 Croonian Lectures at the Royal College of Physicians (Garrod, 1908) in which he described four inborn errors of metabolism: alkaptonuria, albinism, cystinuria, and pentosuria; porphyria was to follow later. As the subject of his last scientific paper published in the *Quarterly Journal of Medicine* in 1936, the proof of this postscript on porphyria was not corrected because its author was too ill (Bearn, 1993). In Garrod's lifetime it is notable that another important disorder of aromatic amino acid metabolism, phenylketonuria, was reported by the Norwegian, Asbjörn Følling in 1934. Garrod was gratified by Følling's discovery and delighted that the so-called imbecility that accompanied the untreated disorder could now largely be prevented by the prompt introduction of a restricted diet in individuals identified as a result of biochemical screening in the pre-symptomatic neonatal period.

Mendelian Disorders in Man

Inspection of the extraordinary catalogue of Mendelian characters in humans, Mendelian Inheritance in Man, compiled by Victor McKusick at Johns Hopkins at present reveals 9,654 established gene loci of which 7,531 have chromosomal assignments*. Of these, more than 9,000 are of autosomal traits: 500 are X-linked and 38 Y-linked. Of note, 37 mitochondrial loci are also in the catalogue. This rapid expansion of genetic information in relation to man is an apotheosis of Garrod's initiating studies and represents the fruit of clinical observation combined with biochemical genetics.

It is often stated that Mendelian disorders are individually rare, but as a group they represent an enormous burden of illness. Even in Western societies, where consanguineous marriages are the exception rather than the rule, the overall birth frequency estimated at several centres in Europe and North America is approximately 1%. Dominant disorders occur with an estimated frequency of 7 per 1,000 live births, recessive disorders occur with an estimated frequency of 2.5 per 1,000 live births and X-linked disorders, excluding colour blindness, have an estimated frequency of approximately 0.4 per 1,000 live births. Between 6% and 8% of children admitted to hospital are estimated to have Mendelian disorders.

Scientific Advances in Human Genetics

In considering the many technical developments in human genetics from 1950, it is salutary to consider the history of alkaptonuria research itself. From Garrod's early report in 1899 and his recognition of consanguinity in the parents of affected subjects in 1901, he developed the concept put forward in the lectures entitled "Inborn Errors of Metabolism" and "The Study of Chemical Individuality" – subjects he returned to repeatedly during his professional life. In 1958, Garrod's deduction that there was a defect in the enzymatic cleavage of the benzene ring in aromatic amino acids was brilliantly confirmed. In two *post mortem* liver samples, La Du in 1958 showed a marked deficiency of homogentisate 1,2 dioxygenase activity a half century after Garrod had delivered his seminal Croonian Lectures.

It took until 1993-1994 for the human locus of alkaptonuria to be assigned. This was achieved by homozygosity mapping based on the occurrence of consanguinity in the parents of alkaptonuric subjects; the long arm of

---

* Online Mendelian Inheritance in Man (OMIM), National Center for Biotechnology Information. Available at: http://www.ncbi.nlm.wih.gov/omim.

chromosome 3 was identified as the locus for this trait. In 1995, a fungal homologue of homogentisate 1,2 dioxygenase was cloned from *Aspergillus nidulans* and was later used in 1996 by Fernandez-Canon and colleagues to identify a cDNA encoding the human protein from a search of an expressed sequence tag library. The identity of the clone was confirmed by functional expression of the active enzyme as a glutathione S transferase-fusion protein in *Escherichia coli*. Later, genomic clones that encode the enzyme were identified and assembled from an EMBL 3 λ-library and fluorescent *in situ* hybridisation refined the mapping of homogentisate 1,2 dioxygenase to the 3q21-23 region. The first mutations in the coding region of this gene were identified by sequence analysis in 1996, revealing a C->T transmission at codon 230 and a T->G transversion at position 1028, respectively, responsible for the P208 and V308G mutations in the homogentisate dioxygenase protein. Expression of the variant enzymes as fusion proteins in *E. coli* later confirmed that these mutations caused loss of catalytic function of the enzyme, thus completing the first steps in the analysis of the *molecular* genetics of alkaptonuria late in the twentieth century.

Technical Advances as Applied to Human Genetics

This catalogue of achievements appears deceptively simple, but it has been based on astonishing progress in the technology of biological science since 1950. Molecular analysis of human genes would have been unconscionable before the deduction in 1953 of the structure of DNA by Watson and Crick. After all, it was not until the studies by Avery and Griffiths and molecular phage genetics that DNA, indeed, was proven as the hereditary material. The study of Mendelian disorders now benefits from a host of technological discoveries and applications to biology, including amino acid analysis, peptide fingerprinting (used to great effect by Vernon Ingram in 1956 to identify the point mutation in sickle cell haemoglobin), the identification of mRNA and the genetic code in the 1960s and the use of DNA and RNA hybridisation techniques for analysis. Other technical discoveries amongst a long list of important applications in human Mendelian genetics have been the use of restriction endonucleases, the cloning of recombinant DNA and Southern blot hybridisation. The introduction of DNA sequencing by Sanger, of site-directed mutagenesis and recombinant protein expression, as well as synthesis of artificial oligonucleotides combined with the polymerase chain reaction greatly accelerated our capacity to isolate sections of our genome: the sequences can be amplified for detailed examination – a technology that has contributed enormously to the definitive diagnosis of Mendelian disorders.

Linkage analysis has been critical for the identification of human disease traits and their localisation within the genome. Now, the emerging draft of the human genome sequence, published in *Nature* and *Science* in 2001, has also accelerated our capacity to identify genes implicated in human diseases that map within candidate regions identified by linkage analysis. Latterly, the introduction of mouse transgenesis, particularly the use of embryonic stem cells for mouse gene knockout technology, as well as sophisticated methods for chromosome sorting and molecular cytogenetics has accelerated understanding of the molecular pathogenesis of inherited traits. Now, authentic animal models of human diseases that are not otherwise easily susceptible to study can almost be produced at will. There is already a long list of disease genes that have been cloned positionally using the draft genome sequence. Since the estimated number of human genes is, say, between 35,000 and 50,000, it is easy to see that even McKusick's Online Mendelian Inheritance in Man (see Footnote on page 106) is, at present, an incomplete catalogue.

From this rapid review of progress, it is apparent that the study of Mendelian disorders in man remains a burgeoning field with advances in many different aspects. Garrod's original biochemical genetics is an established discipline worldwide. In most large hospitals and regional referral centres, neonatal screening for treatable errors of metabolism such as hypothyroidism and phenylketonuria is conducted. Reproducible procedures to measure the activity of complex enzyme systems are available for the diagnosis of rare metabolic errors, and already recombinant proteins are available to repair inherited deficiencies. Some of these proteins, such as recombinant human glucocerebrosidase (imiglucerase) used for the delivery of a key lysosomal enzyme involved in glycolipid breakdown, is already a world best-selling drug for the treatment of Gaucher's disease. As a result, there is a competitive market to generate suitably targeted proteins to correct other lysosomal enzyme deficiencies. In 1949, Linus Pauling examined the physicochemical characteristics of the drepanocyte of sickle cell anaemia in which tissue injury results from the formation of paracrystalline tactoids of sickle haemoglobin after deoxygenation. Pauling rightly coined the term "molecular disease" for sickle cell anaemia to indicate that mutations in any protein may contribute ultimately to pathology as a result of molecular aberrations. Pauling's dictum was neatly followed by the studies of Vernon Ingram in Cambridge who showed that the diverse haematological and clinical manifestations of sickle cell anaemia resulted from a single amino acid change at position 6 of the human β-globin chain. Ingram's findings ably supported Pauling's concept of molecular disease and immediately identified the one-to-one relationship between the genetic trait and the qualitative change in protein structure and function.

Morbid Anatomy of the Human Genome

Victor McKusick has been credited with the concept of the morbid anatomy of the human genome modelled on earlier studies of anatomy and pathology that has characterised many hundreds of years of medical instruction. McKusick's catalogue, now as the Online Mendelian Inheritance in Man on the World Wide Web (OMIM), is a formidable resource for the description of human Mendelian traits; it provides information on the molecular analysis of mutant proteins and gene mapping. The human genome, of course, has already yielded targets for treatment, hence the development of therapeutic proteins and the potential for gene therapy to correct the manifestations of disease. As with conventional anatomy, physiology and pathology, there is an almost seamless link between McKusick's morbid anatomy catalogue and studies of the molecular pathophysiology of human disease.

Molecular Pathophysiology

Sickle cell anaemia itself provides a vivid example of the sterility of molecular genetics alone in relation to human disease. The diverse manifestations of sickle cell anaemia with its chest syndromes, painful bone crises and the effects of secondary hyposplenism, as well as neurological manifestation, gain little in understanding from knowledge of a single amino acid change. For many years, the treatment of sickle cell anaemia has been unsatisfactory and palliative ,involving blood transfusions and the simple secondary measures of infection control. Latterly, there has been further understanding of genetic co-factors that contribute to the severity of sickle cell disease, such as persistent expression of foetal haemoglobin which ameliorates clinical expression of the disease. Recently, it has been possible to improve the manifestations of sickle cell disease and reduce the number of attacks of those at risk by modifying the expression of foetal haemoglobin with the use of hydroxyurea therapy, which modifies the maturation profile of erythroid cells from the bone marrow.

The study of molecular pathophysiology is now an appropriate pre-occupation of medical science involving the free study of transgenic animal models of disease, *in vivo* studies of mutant proteins as a result of site-directed mutagenesis and recombinant protein expression – often followed by intricate structure-function studies using crystallography. There have been some spectacular successes in this work in the understanding, for example, of the molecular basis of α_1-antitrypsin deficiency and of a rare dementia, both of which are due to the aggregation of serpin molecules by a mechanism that has been identified by solving the three-dimensional structure of their cognate serpin monomers. An understanding of the intramolecular arrangements and interactions that are responsible for the serpin polymerisation immediately suggest small molecules that can be modelled therapeutically to inhibit the specific molecular pathology.

"The Debt of Science to Medicine"

Studies of molecular pathophysiology have similarly been extended to patch clamp analysis of mutant membrane channels involved in the transport of ions in the inherited channelopathies affecting muscles, nerves and cardiac tissue. Further investigations have also led to the identification of diverse membrane transporters for sodium, chloride, potassium and hydrogen ions implicated in renal tubular acidosis and disorders of mineral balance. Mutations in these transporters contribute to many newly identified syndromes associated with hypertension as well as salt and water homeostasis. Isolation of the gene that encodes the cystic fibrosis transmembrane regulator (CFTR), a chloride channel has been a signal achievement; the CFTR was the first human member of a large family of ATP-transporters to be identified. The identification of biological signalling pathways in molecular cell biology also holds much promise for the understanding of molecular pathophysiology of several unexplained Mendelian disorders and vice versa. The tumour-suppresser gene PTEN, implicated in the hereditary cancer disease, Cowden syndrome and spontaneous tumours of the prostate and breast, as well as gliomas, affects signalling by phosphatidylinositol 3, 4, 5 triphosphate and is an important example. We must remember that it was Garrod in his Harveian Oration at the Royal College of Physicians in 1924 on the subject of the "Debt of Science to Medicine" who perhaps first quoted the now familiar letter by William Harvey at the end of his life stating:

> Nature is nowhere accustomed more openly to display her secret mysteries than in cases where she shows traces of her workings apart from the beaten path; nor is there any better way to advance the proper practice of medicine than to give our minds to the discovery of the usual Law of Nature, by careful investigation of cases of rarer forms of disease. For it has been found, in almost all things, that what they contain of useful or application is hardly perceived unless we are deprived of them or they become deranged in some way. (Garrod, 1924; Willis, 1848)

The Influence of William Bateson on Human (Biochemical) Genetics

Garrod did not work alone, and his ideas about genetics and evolution based on selection leading to human phenotypes were cross-fertilised by the work of the brilliant and heretical biologist William Bateson, about whom we have heard so much in Professor Patrick Bateson's Darwin Lecture in Chapter 5. A review of William Bateson's work identifies three brilliant discoveries, all of which continue to have application in human Mendelian disorders. Bateson worked with several almost equally talented colleagues, most of whom were women who attended Newnham College, Cambridge. In 1902 he identified the phenomenon that he called *"coupling"* – genetic linkage. The discovery followed investigations into segregation of flower colour and pollen structure in a distantly related member of the legume family of which Mendel's garden pea, *Pisum sativum* is a member, the sweet pea. Clearly the phenomenon of genetic linkage in part violates the Mendelian principle of independent assortment, and it is a remarkable fact that virtually no linkage would have been detected in even a larger series of experiments had Mendel conducted them in the seven characters that he studied in *P. sativum*. Notwithstanding Bateson's strangely late and almost reluctant conviction that linkage provided confirmatory evidence of the occurrence of chromosomal inheritance, the phenomenon of genetic linkage has been critical for mapping disease genes. Assignment of a chromosomal locus has been critical for isolating numerous "disease genes" by molecular analysis of human DNA.

William Bateson also described *epistasis* – the influence of two or more genes on expression of the same trait. Much of his research was carried out on floral pigments, but also included farmyard animals such as the domestic hen in which he studied the inheritance of coat or feather colour.

The understanding of epistatic inheritance provided the theoretical underpinning of Garrod's later work, i.e., the concept of diathesis and pre-disposition to disease, as a result of interactions between genes leading to a complex phenotype. With R.C. Punnett, Muriel Onslow and Rose Scott-Moncrieff, Bateson contributed to the understanding of genetic complementation in complex biochemical pathways. The biosynthesis of pigments was later raised to a high experimental level by Beadle and Tatum in the study of the quinones responsible for eye colour in *Drosophila* and in the study of auxotrophic mutants in the mould, *Neurospora*. (Beadle, 1958)

One of Bateson's most important discoveries was that of *homeosis* or *homeotic variation* in his study entitled *Materials for the Study of Variation* in 1894. His early work in this regard was based on studies of plants, for example, common doubling mutants that have been well systematised in plant teratology. Bateson originally considered that homeotic mutants always involved the transformation of one part into the likeness of another to which it is related in terms of development. In reviewing these abnormalities in humans, he was fascinated by autosomal dominant traits affecting development of the digits such as brachydactyly, polydactyly and syndactyly. Brachydactyly is caused by the reduction in phalangeal articulations in the fingers and toes, as shown by early X-rays by Drinkwater. Farabee had also reported that many brachydactylous subjects are themselves of unusually short stature. *Homeotic genes* are now critical components in our understanding of the development of many organisms from the worm *Caenorhabditis* through to *Drosophila* and from early chordates such as *Amphioxus* to higher vertebrates, including man. Homeotic genes are defined as those in which mutation results in the transformation of body parts in structures normally found elsewhere. When Herman Müller studied X-ray-induced mutants of *Drosophila,* transformations of this kind were very frequent; it was later shown in the fly that many of the homeotic genes lie in two clusters, the bi-thorax and antennapedia complexes. Wild-type alleles of these genes are responsible for specifying the segmental identities of developing organisms, and in *Drosophila* they are expressed in overlapping domains along the antero-posterior axis of the embryo. The products of homeotic genes are proteins containing *homeodomains*, which serve as transcriptional regulators. It is now known that in humans and mice homologous genes (*Hox* genes) exist; mutations in these have lately been shown to cause homeotic transformations in vertebrates, including man.

At least 40 mammalian *Hox* genes have been identified which map to at least 9 clusters that possess the same transcriptional orientation. Detailed analysis of the gne sequences show that they can be classified into at least 13 parologous sub-groups, the majority of which contain one gene from each cluster suggesting that they arise in evolution by successive duplication of an ancestral gene group. The multi-gene family of lomeobox genes are characterised by the presence of a semi-conserved sequence of 180 nucleotides found within the coding region. This homeobox sequence encodes a protein motif of 60 amino acids, the homeodomain which is a key component of the homeoprotein or homeodomain protein. Although conservation of the homeodomain sequence is not complete, the definition of the sequence is usually based on 12 highly conserved residues with an invariant predicted secondary structure. For the most part, homeodomain proteins prove to be transcription factors that recognise site-specific DNA sequences. Binding to DNA is mediated by the homeodomain as a result of its conserved 3-α-helix structure, the second and third helices of which form a helix-turn-helix motif, allowing residues of the third homeodomain helix to make contacts with bases in the major group of

DNA. Some homeodomain proteins possess two different DNA binding domains with a subsidiary zinc finger motif or a paired domain to accompany the homeodomains.

Genetic studies in flies and nematodes, as well as the study of natural mutants of vertebrates and man, indicate that homeobox genes posess a pluripotential role in the control of development. These genes determine spatial patterning, cell fate and cell differentiation by regulating a vast array of other regulatory genes and target genes. Thus, tight control of transcription leads to segmental patterning and temporal flow of differentiation information during embryonic development. Most homeobox genes in mammals have been isolated by searching for conserved sequences present also in *Drosophila* and *Caenorhabditis*. Hox cluster genes are expressed in overlapping antero-posterior domains of the embryonic mesoderm and neural crest derivatives in the developing brain. Their sequences interact with soluble factors and other gene products, including retinoic acid, that regulate positional signals within vertebrate embryos. The function of some mammalian *Hox* genes has been investigated by means of antibodies to homeodomain peptides in embryonic animals and by experimental gene disruption by homologous recombination in embryonic stem cells to generate mice harbouring specific homeobox mutations.

Human Developmental Genetics

For many physicians, the area of Mendelian genetics in man appeared to be restricted to simple phenotypes such as those induced by mutations in structural proteins such as collagen, fibrillin (e.g., *osteitis fragilitas* Marfan's syndrome) or enzyme deficiencies. However, rapid translation of the developmental genetics of plants and animals into higher vertebrates and man has greatly extended our understanding of the control of human development and to the molecular pathogenesis of congenital anomalies.

Recently, a group from China has studied one particular form of brachydactyly associated with dwarfism and the radiological abnormalities known to Bateson. Gao et al have shown this year that point mutations in residues common to a group of signalling proteins of the so-called *hedgehog* family in vertebrates and some invertebrates are responsible for this condition. *Hedgehog* are segment-polarity genes that encode morphogens in vertebrates, in which they are secreted by cells on either side of segment boundaries during development. Hedgehog proteins bind to receptor proteins and induce expression of wingless, patched and decapentaplegia genes, as well as phosphorylation of the protein, fused. Decapentaplegia is related to the transforming growth factor-β (TGF-β) and bone morphogenic protein, type 4 (BMP-4).

The brachydactyly-associated mutations in *Indian hedgehog* disrupt the regulation of hedgehog signalling during limb development. Mutations in the cleaved N-terminal domain of the Indian hedgehog protein disrupt its interactions with cholesterol following autoprocessing and cleavage, as well as binding of the processed polypeptide to another developmental protein, known as *patched*. The normal interaction between Indian hedgehog and patched inhibits the action of patched on another developmental gene product, *smoothened.* These interactions in the hedgehog pathway normally promote downstream transcriptional signalling to the TGF-β and BMP-4 proteins. In mice, expression of *Indian hedgehog* is essential for chondrocyte proliferation, and embryos lacking exon 1 of this gene have foreshortened limbs and unsegmented, uncalcified digits. Modelling the structure of the N-terminus of human *Indian hedgehog* protein Inn suggests that the point mutations identified in brachydactyly type A-1 would interfere with Inn binding to the patched receptor.

An even more vivid example of a single gene defect responsible for a complex developmental condition is provided by the genetically hereogeneous disorder, holoprosencephaly. Holoprosencephaly may arise spontaneously, but the condition is influenced by environmental factors, including maternal diabetes which gives a 200-fold increased birth risk. Holoprosencephaly is associated with variable midline abnormalities including cyclops, median cleft lip and palate, hypoteleorism and pituitary deficiency. In some severely affected infants, there is a defective nasal septum and agenesis of the *corpus callosum*. One form includes the Smith-Lemli-Opitz syndrome with cardiac abnormalities such as atrial septal defect and arrythmias. Holoprosencephaly occurs in 1 in 200 spontaneous abortions and up to 1 in 1,600 live births. There appear to be at least five genetic syndromes which map independently: Smith-Lemli-Opitz syndrome is caused by deficiency of 7-dehydrocholesterol reductase; the holoprozencephaly type 3, which maps separately to chromosome 7q36, is caused by mutations in the sonic hedgehog protein. Other forms of holoprosencephaly include HPE 4 that maps to chromosome 18 in which mutations in TIGF, a gene encoding a homeo-protein that interacts with retinoic acid receptors, interfere with TGFβ signalling. Holoprosencephaly type 5 maps to chromosome 13 and is caused by mutations in a zinc finger protein, homologous to the homeotic gene that interacts with hedgehog to maintain parasagittal identity, *odd-paired wingless.* The human homologue of this gene is also mutated in *situs inversus.*

Not only do these studies in holoprosencephaly demonstrate that a single gene influences the coordinated expression profiles associated with complexities of facial and midline human development, but they identify epistatic factors in the pathway. Thus, the interaction between cholesterol and the hedgehog signalling pathway is immediately reflected by the effects of deficiency of 7-dehydrocholesterol reductase on the downstream action of the sonic hedgehog gene product in *HPE 3.* At the same time, studies of these horrifying congenital abnormalities have provided essential information about human development; they build directly on the early identification of epistasis and homeosis by Bateson.

Perhaps the most striking revelation in this field, hitherto, has been the recognition that subtle genetic influences may influence a uniquely human characteristic, that of speech and language. Recently Professor Monaco's group in Oxford has identified the genetic basis of a unique speech and language disorder, SPCH 1 (OMIM – footnote I). SPCH 1 is characterised by abnormal facial movements that affect the articulation of words, defects in language processing and the break-up of words into their individual phonemes; there are also defects in grammar usage with failure to comprehend the inflections and syntactical references of everyday speech. These latter defects appear to be remediable to some extent by speech therapy. An unusual feature is that individuals affected by the speech and language disorder in the large pedigree described have normal or near-normal non-verbal skill and intelligence; apart from the language defect, they appear to be quite healthy. As a result of studying another single individual with identical specific speech and language disorder who had a balanced translocation in the long arm of chromosome 7, Lai et al. (2001) were able to map and identify the gene *FOXP 2* that encodes a forkhead-domain (*winged helix*) transcription factor. In the original SPCH 1 family, a single point mutation, R553, was found in the *FOXP 2* gene in all affected individuals. This missense mutation, in which a histidine replaces an arganine codon 553, affects a conserved domain in all forkhead-transcription proteins that is involved in homeodomain recognition. Studies in embryonic mice show that *FOXP 2* is expressed in specific areas of the developing brain, especially in the neopallial cortex.

Other human FOX genes have been implicated in an inherited glaucoma syndrome, in thyroid agenesis and in distichiasis, a congenital abnormality of eyelash development. Distichiasis is also an inherited homeotic variant recorded by Bateson in 1894. Clearly, the studies of SPCH 1 offer a fascinating entry into the science of functional neural development relating to human language. Whilst it is uncertain as to whether polymorphic variation in this gene will explain dyslexia and other dyspractic syndromes in human development, the identification of genetic determinant of speech and language in man surely registers the birth of the new science of human cognitive genetics.

There is not space here to survey other aspects of Mendelian inheritance in man. They are to be covered by Dr. Knudson (Chapter 11: Human Cancer Genetics) and Dr. Lucio Luzzatto (Chapter 7: Malaria and Darwinian Selection in Human Populations). Rather than simply produce a catalogue of the many fascinating monogenic disorders that constitute part of modern human Mendelian genetics, it might be preferable to ask what we can learn from the study of these conditions. Clearly Mendelian disorders of man have revealed diverse mechanisms of human mutations, including the so-called triplet repeat diseases. This mechanism operates in several neuro-degenerative disorders (e.g., the CAG polyglutamine encoding repeats in Huntington's chorea), which as dynamic mutations contribute to the phenomenon of genetic anticipation in these syndromes. Mendelian disorders have also shed light on novel components of disease pathways, which will come to assume greater importance as DNA-based reagents for diagnosis and ultimately for treatment emerge. Since many Mendelian traits are not yet susceptible to specific treatments and are associated with considerable human misery, their study so far has yielded only improved capacity for diagnosis; in particular pre-natal diagnosis in at-risk pregnancies using amniocentesis, chorionic villus sampling or, it is to be hoped, reliable analysis of the target loci obtained from foetal DNA within the maternal circulation. There have, however, been some spectacular revelations from the identification of the causes of Mendelian diseases including cystic fibrosis. Cystic fibrosis genetics rather than endless biochemical study yielded the underlying cause of the condition; understanding the function of the novel chloride channel transporter will ultimately be the barometer by which our understanding of the molecular pathogenesis of cystic fibrosis will be assessed. Finally, I hope I have illustrated here that Mendelian disorders in man can now be implicated in very complex phenotypes that affect human development and cognitive abilities, including language and behaviour.

Interactions between Human Genes and the Environment

William Bateson and Archibald Garrod, as Professor Bateson has told us, were, at heart, interested in human evolution and it is in the study of gene-environment interactions relating genetic variation to human selection that we most closely approach their area of interest. It is after all only in terms of evolution that biological observations make sense. In humans, this reflects our historical past and our interactions between our environment and other individuals. Of course, our evolved constitution and genetic programme carries with it our species's investment for the future. Perhaps the most vivid example of Mendelian defect and environmental interactions is provided by the well-studied sickle cell anaemia, to which Dr. Lucio Luzzatto refers in Chapter 7. The evidence for a balanced polymorphism in relation to sickle cell anaemia is very strong. It is also notable that it was the research into sickle haemoglobin by Vernon Ingram at the suggestion of Francis Crick in Cambridge that revealed the cause of the electrophoretic differences between sickle haemoglobin and normal adult haemoglobin. The immediate implication of Ingram's work was that there was a direct relationship between a genetic mutation and the sequence of a human protein.

Garrod was fascinated by the chemistry of pigments and this perhaps explains why albinism was one of his early inborn errors. It is perhaps not fortuitous that albinism also affects the pathway of tyrosine metabolism. Many types of albinism are associated with abnormalities of tyrosinase - the enzyme which converts tyrosine to di-hydroxyphenylalanine and which catalyses the formation of di-hydroxyphenylalaninequinone as key initial steps in the formation of eumelanins. Thus, deficiency of tyrosinase leads to albinism. A striking example of a gene-environment interaction in this pathway is provided by the Siamese cat and the Himalayan mouse, which show residual pigmentation in peripheral coat hair and skin cooled by the environment due to the presence of temperature-sensitive variants of tyrosinase. A similar mutation is seen in their human counterpart with partial albinism but melanin formation in distal body hair and in the eyebrows due to the presence of a temperature-sensitive mutant of human tyrosinase. It is difficult to imagine the temperature-sensitive mutant as a positive selective force in cold areas, but in the wild, animals with temperature-sensitive albinism would carry a selective advantage in desert regions for protective coloration in relation to sand or light-coloured rock.

There are other dramatic examples of metabolism of adverse interactions between specific environments and genetic constitution. One notable example is that of hereditary fructose intolerance, which is due to mutations in a tissue-specific isozyme of aldolase. Aldolase B has preferential activity for the metabolic incorporation of fructose and the related sugars, sucrose and sorbitol; deficiency of aldolase B causes the disease. Patients with hereditary fructose intolerance remain perfectly well and can stand starvation provided they avoid these sugars; breast milk is harmless, but on weaning and exposure to fruit sugars (which are present almost ubiquitously in modern foods) severe metabolic disturbances, accompanied by pain and hypoglycaemia, ensues. The biochemistry of the fructose-induced disturbance of metabolism is complex, but it is notable that those tissues such as the renal cortex, intestine and liver that are responsible for the metabolic incorporation of exogenous fructose are those which may incur fatal injury on continued exposure to this noxious sugar. Children that survive the stormy period of weaning develop marked taste aversions to the foods that provoke the symptoms and are often referred because of psychological difficulties associated with food fads – and rejection of the maternal figure is occasionally suspected. Adults and children with hereditary fructose intolerance have a striking absence of dental caries, indicating a modification of eating habits and – if it were needed – powerful evidence of the adverse effect of dietary sugar on the development of dental caries. Many deaths due to the inadvertent use of fructose in otherwise healthy individuals with mutations in the aldolase B gene have been recorded, most notably in Germany where a retrospective diagnosis of death by molecular analysis of aldolase B genes obtained from small biopsy samples of liver or by molecular analysis of aldolase B genes in close family members have been reported.

Hereditary fructose intolerance represents a vivid example of gene-environment interactions. Population studies show a high frequency in Britain of the most common mutant allele of aldolase B – sufficient to give a heterozygous frequency of about 1.4%. Since, like phenylketonuria, hereditary fructose intolerance is an entirely preventable nutritional disease with a pre-symptomatic period in the neonatal life, there is a strong case for the introduction of genetic screening based on blood samples universally collected within the first week of life on Guthrie cards (Ali, Rellos and Cox, 1998). Hitherto, attempts to introduce a pilot scheme to investigate the practicability and outcome of such a programme have been rejected.

Biological Aspects of Population Genetics: Nutritional Influences

Several studies have confirmed the widespread distribution of some mutant alleles of aldolase B within the populations of European descent. The most frequent mutation designated by the single amino acid letter code, A149P, prevalent in Northern Europe studied using an intragenic polymorphism has shown that, in all probability, it arose on a single ancestral haplotype and spread, presumably by genetic drift. Another mutation, A174D, is found in Southern Europe in a different distribution. It is by no means certain that these mutations have arisen by genetic drift alone, and it remains formally possible that selection has occurred.

Several mutations of aldolase B have a widespread distribution, at least in Europe, and, as with the β-globin E6V mutation responsible for sickle cell anaemia, fructose intolerance may have been selected on the basis of a positive selected advantage. The introduction of fructose and sucrose into the diet has been a very recent phenomenon in human history and is based principally on the industrialised extraction of sugar from cane and only latterly beet. Much of the sugar industry was based on European slavery which expanded in the seventeenth century. Current sugar consumption involves between 60 and 80 grams of fructose equivalence daily, but before 1700 the mean annual consumption of sucrose was less than 3.7 kg per capita. Slavery to sugar has been responsible for many human ills, and indeed led to the enhanced movement of genes in several populations (Cox, 2002).

One particular nutritional disease, hereditary haemochromatosis, provides an analogous example of the effect of an inherited trait and interactions with environmental factors in the genesis of disease. Adult haemochromatosis is a familiar but uncommon nutritional syndrome due to the deposition of iron within parenchymal such as the liver, anterior pituitary, β-cells of the pancreatic islet and the myocardium. Haemochromatosis characteristically presents in middle-aged men with a combination of impotence, liver disease, arthropathy with or without diabetes and, occasionally, cardiac arrythmias. Liver biopsy shows a pigmentary cirrhosis without prominent cell death; individual cases respond well to the removal of iron by phlebotomy. In the full-blown case, between 15 and 50 grams of storage iron is present in the tissues and irreversible hepatic and cardiac injury is present. It is believed that removal of iron before cirrhosis and end-organ failure of the endocrine system develops is associated with a normal or near-normal life expectancy. However, established iron storage disease in adult haemochromatosis shortens life, and a high proportion of deaths from the condition are due to cirrhosis with or without complicating hepatocellular carcinoma and the consequences of diabetes; established haemochromatosis is associated with a reduced quality of life as a result of hypogonadism and joint disease. Most patients with full-blown clinical haemochromatosis are men who drink alcohol, although it is not understood how alcohol interacts with the product of the predisposing recessive allele of the HFE gene to cause this condition.

Haemochromatosis is an uncommon widespread disorder in European populations and occurs in the presence of one or two common haplotypes reflected in the linkage disequilibrium of HLA class I alleles (typically, A3, B7 or B14 DR3). As a result of formidable research undertaken by the erstwhile Mercator Genetics Company (and at least 20 years after the linkage association of adult haemochromatosis with MHC class I loci on the short arm of chromosome 6) point mutations in a gene termed *HFE* were shown to predispose to the condition. Most patients with the disease in Northern Europe are homozygotes for a missense mutation, termed C282Y, resulting in a replacement of a conserved cysteine residue in a non-classical class I molecule by a tyrosine. This mutation disrupts the interaction with the β-2 microglobulin gene, thereby abrogating co-translational processing of the protein during biosynthesis and inhibiting cell-surface expression of the mature *HFE* gene product. Adult haemochromatosis is caused by a persistently increased avidity for nutritional iron leading to the storage of toxic iron in the tissues.

Studies of the frequency of the C282Y allele show that it has a high frequency (approximately 10%) in the Northern European population within the penumbra of territories subjected to Viking invasions and is distributed widely. Given that this allele, which predisposes to haemochromatosis, has a population frequency that is much greater than that which would be expected by recurrent mutation, it may have arisen as a result of genetic selection. The selective factor that operates on the haemochromatosis locus at human chromosome 6p is unknown, but an obvious suggestion is that it provides advantageous protection against iron deficiency. Iron deficiency is the most common organic disease of mankind, and it is estimated to occur with a frequency of approximately 30% in the world's population. Possession of one or even two C282Y alleles of HFE that predispose to iron loading may confer selective advantage under conditions of reduced iron availability. Those most at risk from iron deficiency today include premature infants, young children, mothers and the elderly within poor socio-economic groups. Iron deficiency is no respecter of civilisation and occurs in well-developed as well as poor countries, although its distribution shows variation that is closely correlated with local economic factors and national productivity. We do not know for certain what selective factor operates at the HFE locus. It is certainly surprising that the mutation appears to have arisen only once or twice on a common MHC haplotype and spread within Northern Europe - rather than other parts of the world where, at present, iron deficiency is most prevalent. Since hookworm anaemia is a globally powerful co-factor in the development of severe iron deficiency with impaired work output and productivity, it is surprising that mutations in the HFE gene are not more widely distributed or restricted to those areas that have had a long ancestral history of hookworm infection. Hookworm infestation has been recognised in the Old World and particularly in the Middle East since antiquity.

It is possible that another factor operates to select for the mutations in the HFE gene. One such factor would be microbial infection of the gastrointestinal tract. Most bacteria, including Helicobacter pylori, have an absolute requirement for environmental iron which limits their growth. Fungi, yeast and bacteria have evolved very avid chelators for obtaining ferric iron from the environment (siderophores), which are used to facilitate uptake through a receptor-mediated mechanism. Iron is often limiting for microbial growth, and it seems quite possible that the iron uptake system operating within the small intestinal epithelium would compete effectively for free ferric iron and serve as a selective force against pathogenic microbes including yeast and bacteria such as salmonellae and helicobacter, which otherwise fastidiously conserve this vital nutrient for their own purposes. Ferrireductases are abundant in the intestinal brush border membrane and rapidly reduce soluble complexes of ferric iron in food within the lumen to ferrous iron. Ferrous ions are taken up by the divalent metal transporter DMT 1, also located on the intestinal microvillus membrane. In haemochromatosis, expression and functional activity of DMT 1 is greatly increased, and it is possible that enhanced uptake of ferrous iron by this pathway competes successfully against bacteria, thereby preferentially improving resistance to enteric infection. Since haemochromatosis is a disease expressed principally in middle-aged individuals and is seen only in a proportion of adults

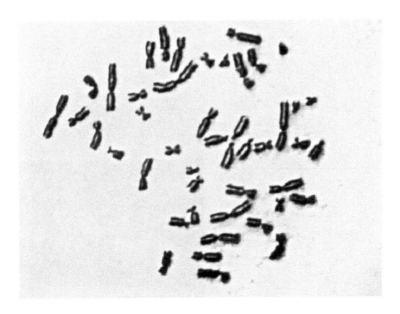

FIGURE 8.1 Human male metaphase prepared from cell culture using hypotonic treatment, acetic-alcohol fixation and drying in air. Aceto-orcein stain.

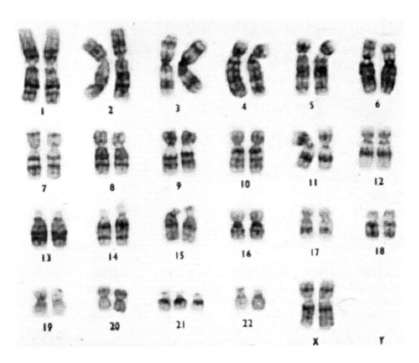

FIGURE 8.2 Karyotype of Down syndrome patient showing trisomy 21, stained by Giemsa banding.

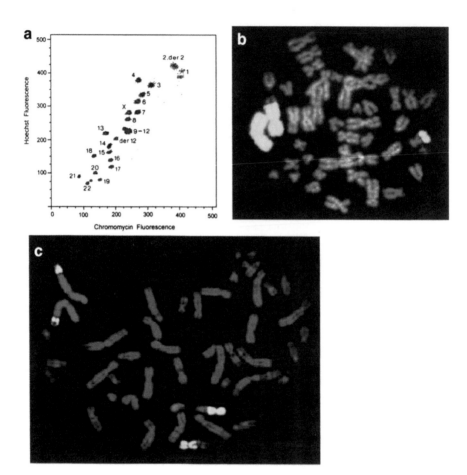

FIGURE 8.3 (a) Flow karyotype of chromosomes from a patient with a *de novo* translocation between the long arms of chromosomes 2 and 12. Note the position of the two derivative chromosomes and that the chromosome 2 derivative sorts with the normal 2.
(b) A metaphase from the above patient to which specific paint probes for chromosome 2
(green) and chromosome 12 (red) have been hybridised to reveal the extent of the translocation.
(c) A normal male metaphase to which paint probes prepared from sorting the derivative
chromosomes in (a) have been hybridised. Note the region of chromosome 12 which is not
painted by probes from both chromosomes. The gap represents that part of chromosome 12
which is deleted in the patient.

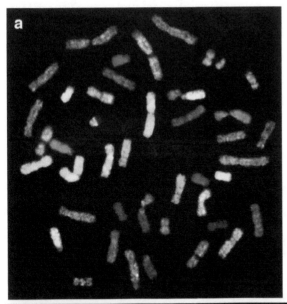

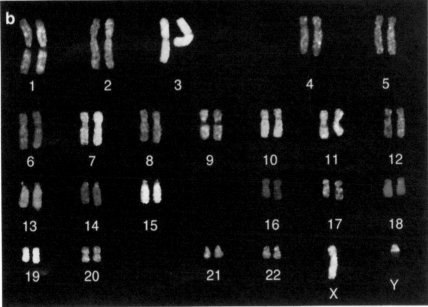

FIGURE 8.4 Multicolour-FISH in which each chromosome pair in the metaphase (a) is identified by a separate colour combination. The karyotype (b) shows the computer classification which automatically gives the correct chromosome number to each member of the pair, together with a distinct pseudocolour.

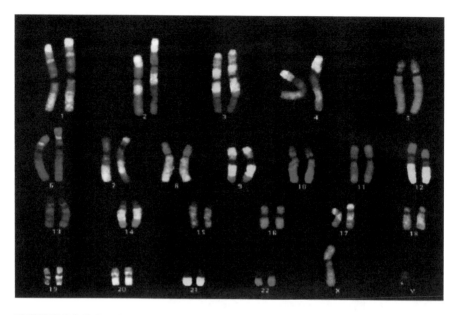

FIGURE 8.5 Colour banding of a human metaphase hybridised with an M-FISH made from gibbon chromosomes. The colour bands represent chromosomal segments which have been rearranged during the divergence of humans and gibbons. Note the complex inversion of chromosome 7 (right).

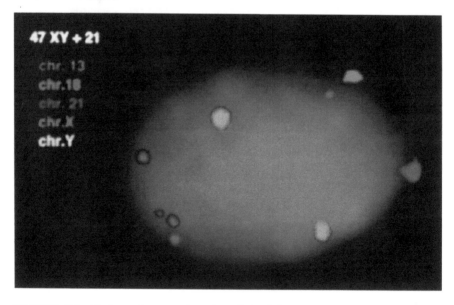

FIGURE 8.6 Nucleus of an uncultured amniotic fluid cell stained with chromosome-specific centromeric (chromosomes X and 18), unique sequence (chromosomes 21 and 13) and Y-chromosome repeat sequence probes. Three chromosome 21 signals (red) indicate that the fetus has trisomy 21, Down syndrome. *Source*: From Divane, A. et al., *Prenatal Diagn.*, 14, 1061–1069, 1994. With permission.

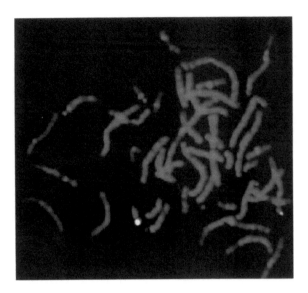

FIGURE 8.7 Metaphase from a patient with a cryptic translocation between the ends of chromosomes 7 and 21. Two closely-linked cosmid markers (one red and one green, but together giving a yellow signal) have been hybridised to the patient's chromosomes. The two cosmid markers have been separated by the translocation which has transferred the green cosmid to chromosome 7 and left the red cosmid on the chromosome 21 derivative; the normal 21 shows a yellow signal indicating the presence of both cosmids.

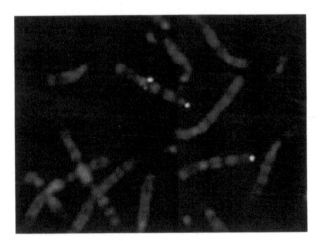

FIGURE 8.8 Williams microdeletion syndrome. Patients with Williams syndrome (supravalvular aortic stenosis, mental retardation and hypercalcaemia) are heterozygous for a deletion of a contiguous series of genes in chromosome 17. This can be identified by a cosmid containing DNA sequence from the elastin gene within the deleted region. Absence of signal from one chromosome 17, as in this cell, is diagnostic. Chromosome 17 is identified by a cosmid clone at the end of the long arm. *Source*: From Connor, M. and Ferguson-Smith, M., *Essential Medical Genetics*, 5th Ed., Blackwell Press, Oxford, UK, 1997, plate 11. With permission.

FIGURE 8.9 DNA-fibre FISH. Three cosmids of 35 kb each from a contiguous sequence in the MHC locus on chromosome 6. *Source*: From Mann, S.M., et al., *Chromosome Res.*, 5, 145–147, 1997. With permission.

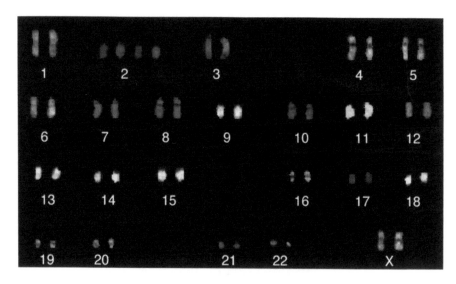

FIGURE 8.10 M-FISH karyotype of orangutan to which human chromosome-specific paints have been hybridised. Numbers refer to the human paints. Note that two orangutan chromosomes are painted by human chromosome-2-specific paint probe.

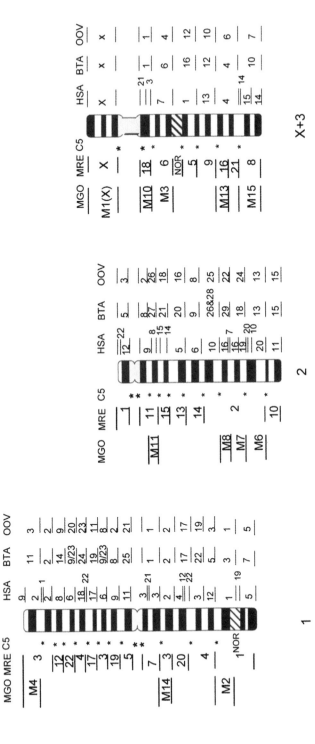

FIGURE 8.11 Idiogram of the female Indian muntjac karyotype showing homologies revealed by chromosome painting with probes from human (HSA), sheep (OOV), bovine (BTA), Chinese muntjac (MRE) and brown brocket deer (MGO). Segments homologous to the extensively mapped human genome are indicated for each species. Asterisks mark sites at which fragments of telomeres have been identified in the Indian muntjac, indicating ancient fusions. *Source:* From Ferguson-Smith, M. et al., *ILAR J.*, 139, 74, 1998. Reprinted with permission from the Institute for Laboratory Animal Research (ILAR), National Academy of Sciences, 500 Fifth Street NW, Washington, DC 20001 (www.national-academies.org/ilar).

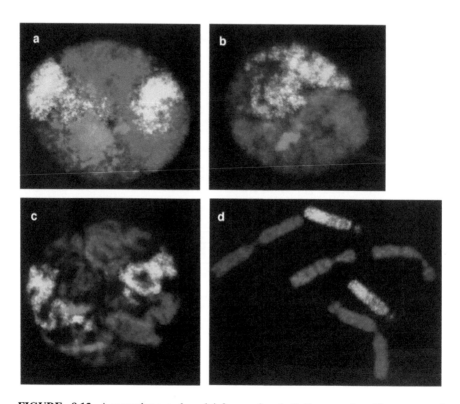

FIGURE 8.12 A metaphase and nuclei from a female Indian muntjac. Chromosome 1 (purple), chromosome 2 (green) and chromosome 3 + X (red) are also shown in prophase and interphase. It is evident that each chromosome occupies a distinct domain, and that chromosome centromeres tend to be clustered in the centre of the nucleus.

homozygous for the predisposing C282Y, allele adverse selection for this genotype would be minimal.

In relation to other gene-environment interactions, there is much more to be said about epigenetic factors that contribute to polygenic disease, which will be discussed by Professor Bell in Chapter 10. We are also to hear more concrete examples of positive selection resulting from genetic variation at the glucose 6-phosphate dehydrogenase locus and in relation to blood disorders such as hereditary ovalocytosis, sickle cell anaemia and β- and α-thalassaemia syndromes which affect the red cell and by unknown mechanisms confer resistance against malaria. In relation to the metabolic syndromes above, more obvious and quantifiable interactions with the environment operate under specific circumstances. These interactions are exactly as predicted by Bateson and Garrod and show that genetic variation serves as a driving force for human evolutionary selection.

Changing Concepts of Human Disease – the Patient as an Evolutionary Product

The last half-century has seen a radical change in our thinking about human diseases. This change owes its origin to the obscure utterances of Garrod, largely viewed as a backroom biochemist with clinical pretensions, and Bateson, an heretical biological figure – as well as an aggressive defender of Mendelism in the face of the forceful mathematical biology that originated with Galton in London. As Professor Bateson pointed out in his Darwin Lecture, William Bateson and Garrod were members of a late-Victorian upper middle-class with strong scientific ideals based on the Baconian philosophy of utility and progress. They were curious and indefatigable experimenters who sought, in their more reflective moments, to identify a unifying biological theme in their discoveries. Bateson was an expansive biologist with a strong and wide-ranging experimental programme including many research collaborators (and some very gifted early women scientists). Garrod, on the other hand, was constrained by his medical perspective. Although Garrod enjoyed the intellectual and technological expertise conferred by his friendship with Gowland Hopkins, as Professor Bateson has told us, because he was a doctor, he tended to adopt what is now termed "the reductionist" approach. Barton Childs in 1989 and Alexander Bearn, Garrod's biographer, have looked for greater insight to Garrod's later masterpiece, *The Inborn Factors in Disease* (1939). In his book Garrod developed the theme that genetic influences and their individual interactions operate in most, if not all, human diseases. This concept, now widely accepted, was founded on Garrod's studies on alkaptonuria; later, he extended his beliefs into the more general theory of biochemical variation and individuality as the basis for, and result of, human evolution.

Garrod himself had a rudimentary understanding of genetics, Mendelian or otherwise; indeed, he was even inferior to Bateson in his ability to understand the contribution of genes quantitatively to the development of a phenotype. Garrod and Bateson, however, both had strong biological propensities in their early life based on the sharp eyesight of the Victorian Natural Historian. Growing up as they did in the period that immediately followed Darwin's death and, in their maturity, the rediscovery of Mendel, which provided the theoretical underpinning for inherited variation, evolution must have been a central aspect of their thinking.

Barton Childs, a distinguished paediatrician at the Johns Hopkins Medical School, has drawn an interesting contrast between Garrod and his predecessors in the development of medical thought and medical training. As it happens, Garrod and Osler each took the post of Regius Professor of Medicine in the University of Oxford, but they represented very different poles of thought. Osler is as revered today as he was then, a man of deep humanity and kindness with a wholesome and full commitment to the study of disease. His influence on medicine was to teach people about disease in all its aspects. In a sense, Osler represented the growth of Socialism and social philosphy following also from the fine Homeric statement: "Nothing human shall be foreign to me". Osler's medicine was compatible with the investment in medical schools and the elevation of medicine to a noble humanity, to be espoused by the middle-classes as a cure for definable ills. The patients presented themselves with a disease that had to be defined by a bacterium or a vascular pathology; they remained under observation until the disease could be treated. Patients were, in a sense, the battleground for attack by the modern army of medicine; the officers were professionals, fully educated in the art of war against microbes, vascular obstruction and malnutrition – in short, pathology. Today, at the beginning of the twenty-first century, the analogy of the patient as a broken machine is still extant and popular. Just as the mechanic can be brought along to fix a loose bearing to prevent overheating, the anti-microbial can be administered to treat the fever. The patient recovers and is discharged for a single, grateful, follow up. Osler's tenets were applauded and the case is closed.

Given the triumphs of the Osler tradition, the influence of Archibald Garrod has been slow to catch on. Garrod as a personality was far less compelling for aspiring clinicians; he was seen as a rather retiring clinician without the theatrical properties of the bedside teacher. He was an old-fashioned backroom boy – the typical butt of English anti-intellectualism. While there is no question that Garrod was well rewarded by his admiring respect for his intellect in his time, Garrod's scientific contributions were only recognised by the cognoscenti and not by most practitioners. I would submit that Garrod's type of medicine has always been regarded as somewhat irrelevant. His fixation with urine pigments at the bench in the face of human illness and suffering on the wards cannot always have earned a sympathetic audience from the clinical medical student anxious for instruction as to how to relieve human ills and, no doubt, how to profit materially by the process. Nevertheless, the intellectual influence of Garrod is still to be found. Alexander Bearn (1993), his biographer, has pointed to the guilt felt by Beadle and Tatum and others on realising that Garrod, through his studies of alkaptonuria, had already set the ground for their "one gene-one enzyme" hypothesis. Beadle and Tatum were awarded the Nobel Prize for their research.

Garrod and his successors in Chemical Pathology were the forerunners of modern biochemical genetics and clinical biochemistry. I believe that Garrod's influence based on evolutionary theory is very much stronger and is a biological concept yet to be fully accepted into medicine. Garrod asked unusual questions: "What disease does this particular human being have?" "In what way does this patient differ from other people, their contemporaries and peers?" and finally, "What can I do to restore this person's unique orientation to the environment?". As Childs has pointed out, Garrod considered human evolution as the focus of his medical thought and he considered each person – as Bateson might have considered each animal – to be an evolutionary product of its time. The patient then could be regarded dispassionately as a less well-adapted product of evolution. Their complaints resulted from an unusual encounter by an unusual individual with environmental factors for which he was uniquely unfit. From his first descrption of alkaptonuria and the inborn errors of metabolism to his later understanding of modern humans as an evolutionary product, Garrod had systematically developed the idea of predisposition to disease.

Returning to the analogy of rare insights from rare diseases, as resurrected by Garrod from Harvey's obscure correspondence, we can see that it was the Mendelian disorders which led Garrod on to the complexities of gene-environment interactions and the concept of multiple genes interacting with external factors to induce a *clinical* phenotype. Interactions with just a few genes are so complex that I might be forgiven for not tackling the difficult area of polygenic inheritance taken up by Professor John Bell. As Professor Bell shows in Chapter 10, huge resources of technology are needed, together with a high level of mathematical analysis for the study of quantitative traits, to bring this exciting field into the area of clinical application and the public health.

It has been said that Osler taught us to practice medicine and that Garrod taught us how to think about it. Although Garrod's understanding of genetics itself was very simplistic, he understood that Mendelian and Darwinian influences through genetic variation are the emerging future of medical thought and practice for the twenty-first century. Beyond the compelling force of bedside teaching from Osler, the truly holistic medicine can only be based on a biological concept of man and the patient as an evolutionary product.

To conclude, I should like to return to the sentimental side of Garrod, the clinical experimentalist. Despite his slightly retiring clinical presence, I believe Garrod was, at heart, at least as humanistic as Osler and other great clinical personalities from Victorian and modern medical schools. In his Harveian lecture given in 1924 at the Royal College of Physicians, entitled: "The Debt of Science to Medicine", Garrod expressed himself with convincing human warmth combined with an academic ideal: "Obviously clinical medicine presents immense fields of scientific research and those who cultivate them have the added satisfaction of knowing that every advance of medical science will, sooner or later, bring in its train some forward movement of the healing art" (Garrod, 1924).

The Modern Academic Clinician

Garrod, unlike the wise and consummate physician, Osler, was at ease also in the laboratory. Although his own publication record consists principally of case reports and didactic clinical essays, he remained, at heart, an experimentalist and a thoughtful biological essayist. His tangible legacy to British medical practice reflects his dedication to creating opportunities for physicians in scientific work. He promoted the idea of a full-time clinical university professor as an investigator who encourages assistants to do research. With Garrod's proposal came the first professional medical unit in a British medical school, equipped with clinical laboratories but with full access to ward in-patients and out-patient clinics; assistant staff were supported by university funds – and all were debarred from the temptations of private clinical work. This model has flourised, and those of us with a leaning towards the application of genetics to the practice and science of medicine find it reassuring that academic clinical ideals in Britain, at least, owe their origin to a Harveian tradition that was, in part, rediscovered by Archibald Garrod. Just as Bateson served as Mendel's Anglo-Saxon exponent, his collaborator, Garrod, was Harvey's advocate; we are their scientific beneficiaries.

References

Ali M, Rellos P, Cox TM (1998) Hereditary fructose intolerance. *J. Med. Genet.*, **35**: 353-365.

Bateson W (1894) *Materials for the Study of Variation.* Cambridge University Press, Cambridge, UK.

Bateson W (1909) *Mendel's Principles of Heredity.* Cambridge University Press, Cambridge, UK.

Beadle GW (1958) *Genes and chemical reactions in neurospora.* Nobel Lectures in Physiology and Medicine (1942-1962), pp. 587-597, Elsevier, Amsterdam.

Bearn AG (1993) *Archibald Garrod and the individuality of man.* Clarendon Press, Oxford.

Dean M (1996) Polarity, proliferation and the *hedgehog* pathway. *Nat. Genet.* **14**: 245-247.

Gao B, Guo J, She C, Shu A, Yang M, Tan Z, Yang X, Guo S, Feng G, He L (2001) Mutations in IHH encoding Indian Hedgehog, cause brachydactyly type A-1. *Nature Genetics* **28**: 386-388.

Garrod AE (1902) The incidence of alkaptonuria: a study in chemical individuality. *Lancet* ii: 1616-1620.

Garrod AE (1908) The Croonian lectures on inborn errors of metabolism. Delivered before the Royal College of Physicians of London on June 18, 23, 25 and 30 1908. *Lancet* ii: 1-7, 73-9, 142-8, 214-20.

Garrod AE (1924) The Harveian Oration on the debt of science to medicine, Royal College of Physicians of London. *B. Med. J.* ii: 747-752.

Garrod A.E. (1931) *The Inborn Factors in Disease: an Essay.* Clarendon Press, Oxford

Griffiths W, Cox TM (2000) Haemochromatosis: novel gene discovery and the molecular pathophysiology of iron metabolism. *Hum. Mol. Genet.* **9**: 2377-2382.

International Human Genome Sequencing Consortium, *Nature* (2001) **409**: 860-921

Lai CS, Fisher SE, Hurst JA, Varga-Khadem F and Monaco AP (2001) A forkhead-domain gene is mutated in a severe speech and language disorder. *Nature* **413**: 519-523.

Roessler E, Belloni E, Gaudenz K, Jay P, Berta P, Scherer SW, Tsui L-C, Muenke M (1996) Mutations in the human Sonic Hedgehog gene cause holoprosencephaly. *Nat. Genet.* **14**: 357-360.

Scriver CR, Childs B (1989) *Garrod's inborn factors in disease,* Oxford University Press, London.

Venter JC, and 279 others (2001) The sequence of the human genome. *Science* **291**:1304-1351.

Willis, R (1848) *The Works of William Harvey (translated).* London: Sydenham Society, p. 616.

10.　　The Genetics of Complex Diseases

John Bell

The characterisation of genetic determinants that contribute to common complex diseases is one of the major opportunities and challenges for modern biomedicine. Genetic contributions to common disease are recognisable in virtually all major diseases, but their precise nature and their relationship to other environmental factors remains, for the most part, obscure. The advance of human molecular genetics, however, has provided an opportunity to characterise the genetic contribution to the disease at the level of individual genes and polymorphisms. It is likely that these genetic data will become increasingly available over the next 10 years and, as such, will have significant implications for the way we classify and treat common human disorders.[1]

Many of the advances in molecular genetics to date have revolved around highly penetrant genetic variants that are responsible for single gene disorders. These include not only rare autosomal dominant and recessive disorders of childhood, but also highly penetrant genetic variants that are recognised to contribute to rare forms of common disease. Most common diseases, including hypertension, diabetes, breast cancer and colon cancer, all have a subset of patients where the disease demonstrates high levels of heritability and where genetic determinants of disease have been attributable to variations of a single genetic locus.[2,3] Despite the fact that these genetic variants contribute only a small amount (usually <5%) to the totality of the clinical disease, in very common diseases, this can still account for a large number of individual patients. For example, the APC locus responsible for susceptibility to colorectal cancer and BRCA1 locus which contributes to susceptibility to breast and ovarian cancer account for a very small fraction of the total disease burden. Nevertheless, they are important and relatively easy tractable genetic causes of common diseases.

Much more difficult, however, has been the identification of genetic determinants of disease which contribute modest amounts of risk and which, because of interactions with other genes or environmental factors, are incompletely penetrant. These loci commonly contribute a relative risk of 3-5 of disease development, and the alleles responsible for disease susceptibility can often be found in high frequency. The characterisation of these loci is, obviously, considerably more complex, although an understanding of these high frequency, low penetrant genetic variants is likely to account for the majority of genetic susceptibility to common disease found in human populations.

Importantly, the manner in which disease susceptibility, arising from highly penetrant single gene disorders and genetic risk, which is contributed to by genetic variants of lower penetrants and higher frequency, will be used in health care is dramatically different. Much confusion has arisen from the assumption that only information from genes of high penetrance will be useful.[4] This arises from the fact that those using genetic information in the past have been used to characterising highly penetrant single gene disorders. The utility of genetic "risk factors" may prove to be considerably more profound than that of the rare single gene forms of disease. There has been considerable debate[1,4] about the value of characterising genetic determinants of common disease susceptibility. Although the benefits of identifying determinants of drug response and metabolism and targets for drug discovery are well recognised, the contribution of disease susceptibility genetics to other aspects of medicine are less well accepted. Recent modelling experiments, however, strongly suggest that even the detection of one half of the genetic risk factors in common disease will have profound implications for risk profiling in populations, greatly exceeding the current power of risk profiling using conventional risk factors.[5] Although these risk factors alone contribute relatively modest amounts to disease susceptibility, stratification of risk may prove to be a crucial health care parameter as the field develops. Physicians, but not clinical geneticists, already use a range of risk factor determinants for making important decisions about large sets of patients. For example, hypertension and hypercholesterolaemia are already both well-recognised risk factors seen in asymptomatic patients that contribute significantly to increased risk of ischaemic cardiovascular disease and stroke. Although these risk factors do not achieve anything near the predictability seen with single gene disorders and, more appropriately, represent the sort of risk seen with low penetrant, high frequency genes in common disease, they are extensively used by physicians to intervene with therapy in individuals and populations at risk of heart attack or stroke. This approach to clinical practice has some limitations but, nevertheless, has proved itself an effective approach for reducing the risk of important causes of morbidity and mortality in individuals at particularly high risk. It is likely that genetic determinants of disease will commonly fall into this risk factor category and, combined with environmental and other genetic determinants, may provide opportunities to further improve our identification of individuals in high-risk categories. As a result, this information is most likely to be utilised and applied, not by clinical geneticists who are likely to retain an interest predominantly in highly penetrant single gene disorders, but by physicians and specialists in the major medical and surgical subspecialties where clinical decisions are routinely taken on the basis of relatively modest risk factors.

Classification of Disease

One of the major impacts of a genetic understanding of disease will be that it allows a classification of disease to develop that is founded in disease mechanisms rather than disease phenotypes. Historically, diseases have been classified predominantly by the phenotype of the patient, i.e., the symptoms and physical signs seen in patients when they present for medical intervention. Although one relates this phenotypic disease classification to practice in medicine at the end of the 19th and early 20th centuries, many of the classifications established over that period are still applicable today. Only in a few cases has the profession moved to a classification system that more accurately represents the pathogenesis of disease and, in the case of infectious diseases, this mechanistic classification has relied heavily on the characterisation of individual microbes and, more recently, their genetic variants that contribute to particular disease syndromes. For most common diseases, this is not yet possible and, although diagnostic processes have moved from pure bedside diagnosis through cellular and biochemical phenotyping, none of this information provides definitive data on the mechanisms that underlie the disease pathogenesis. Genetics has the opportunity to provide that essential link between mechanisms and disease phenotype and has the potential to substantially alter our understanding of disease. Historically, we have relied predominantly on phenotype to define disease entities.

In the 18th century, fever was recognised as a specific disease entity. In a sense, such classification was logical as it was a predominant feature of patients who presented for medical care. Technology grew up around this diagnosis such that temperature could be routinely measured and monitored during the course of the disease, providing opportunities to subclassify and redefine different forms of fever and different types of disease. It was recognised that there were several forms of fever: an intermittent form with fever occurring at different intervals, remittent fever and continuous fever. Books were written, and thoughtful academic monographs provided a correlation between particular types of fever and outcome. In this case, phenotype provided little information about the disease mechanism and the opportunity to characterise disease mechanistically; in many cases a fever had to await the microbiological revolution of the mid- to late-19th century.

Similarly, the classification of diabetes as a disease also fails to recognise the multiplicity of biological mechanisms that can lead to its physiological change. It has only been in the last 20 years that a range of distinct mechanisms, which can all contribute to the crude phenotype of diabetes mellitus, have been identified. The disease still relies on blood sugar measurement as the major diagnostic tool regardless of the multiple pathways that may be disturbed to lead to this outcome.

In the early 1970s, it was recognised that the particular genetic determinants on chromosome 6 within the HLA region determined the susceptibility to a subset of diabetes mellitus found predominantly in children that invariably led to a dependence on insulin for the therapy of the disease. This work was one of the first successful uses of genetics to subdivide a common disease. This form of the disease, now recognised to be associated with HLA genetic variants, represents a disease distinguished as Type I diabetes.[6] This form of the disease is mediated by an autoimmune mechanism that leads to the destruction of the beta cells of the pancreas. Considerably more refinement has occurred around the genetic determinants that are responsible for these effects. The principal genetic variants are now recognised to alter the peptide binding site in HLA DQ molecules, leading to quite dramatic changes in the P9 pocket for bound peptide. This sequence considerably alters the array of peptides that can be bound and hence recognised by T-cells in individuals who are susceptible to the disease. Interestingly, the size and shape of this P9 pocket is determined by residues in the two adjacent alpha helices that surround the pocket and are similar in both man and the mouse strain identified as being susceptible to the autoimmune form of this disease.[7] Other susceptibility determinants, such as the insulin gene, have also been recognised in this form of the disease.

The vast majority of individuals suffering from diabetes mellitus, however, acquire the disease in later life where it can be associated with both obesity and hypertension. The identification of glucokinase as a genetic determinant of this disease was the first clear identification of a disease gene in the maturity onset form of the disease.[8] This enzyme is responsible for the phosphorylation of glucose and, hence, the signalling of beta cells to secrete insulin in response to a glucose challenge. Abnormalities around the binding site for the substrate glucose reduce the enzyme's affinity for its substrate, hence requiring high levels of glucose for the same levels of insulin secretion. Since this original observation, a host of other genetic determinants for Type II diabetes have been identified. A large number of transcription factors have been implicated in disease pathogenesis, including the HNF transcription factors, HNF-1α, HNF-1β and HNF-4α. These regulate gene expression in the B cell, influencing glucose transport, glycolysis and insulin expression.[9] IPF-1 is another transcription factor that regulates pancreatic development,[10] while mutations in neuroD/BETA 2 is another rare cause of MODY diabetes.[11]

Other loci, including the peroxisome proliferator-activated receptor γ(PPAR-8), have been genetically associated with Type I diabetes.[12] Other transcription factors have also been identified from a genetic perspective. More recently, other genes such as the enzyme calpain have been shown to be associated with common forms of the disease,[13] and other sets of genetic linkages have been defined that are likely to lead to the clarification of genetic susceptibility over the coming years.

In diabetes, therefore, led by the original genetic observations around HLA but, more recently, around a clearer understanding of the genetic determinants that contribute to Type II diabetes, it should be possible to develop a classification system that breaks down the clinically heterogeneous population of diabetics into mechanistically defined subsets. This may have important implications in defining appropriate modes of therapy within these subsets and has already provided important information on prognosis and natural history of the various forms of the disease. For the first time, therefore, this may begin to clarify the clinical issues surrounding the clinical heterogeneity of the disease and will allow the reconciliation of this data with mechanistic and genetic information defining disease subsets.

Pharmacogenetics

A second important contribution that is likely to arise from our understanding of genetics is a clear understanding of the variation in response to different therapeutic interventions in patients with common complex disease. It is widely recognised that the individual response to particular therapy varies widely, and it is likely that this represents innate variation in metabolism or in drug response from individual to individual. The study of these variations is referred to as pharmacogenetics and dates back to observations that define variation in metabolism of drug in different patient populations that, in turn, have important consequences both for the therapeutic effect of interventions and for the toxicity associated with blood levels of these agents. Although variation in drug metabolism was widely recognised 25 years ago, its application has not yet come into common practice in a clinical setting. The molecular basis for many of these variations has now been identified with large numbers of polymorphisms in the cytochrome P450 family of enzymes, as well as in acetyl transferases and enzymes such as thyopurine methyltransferase reviewed in Weber.[14] The characterisation of these variants and the application of this to patient populations may allow much improved utilisation of drug therapy, recognising the substantial variation in drug levels that can arise from polymorphisms in drug metabolism.

Similarly, polymorphisms in genes involved in drug action, that is, the pharmacodynamic variations seen in patient populations, may also contribute to more appropriate use of drug therapy. It is already recognised that response to drug classes such as beta agonists or the statins may vary depending on polymorphisms in enzymes associated with response to therapy. The opportunities for improving the use of pharmaceutical interventions by defining at an individual level those likely to respond appropriately, or conversely to develop toxic side-effects, is another significant opportunity that is likely to arise from complex trait genetics. It is self-evident that if common diseases are made up of a set of mechanistically heterogeneous subtypes, then it is likely that individual therapies designed to target individual biochemical pathways are likely to work with varying degrees of success in different subtypes of the disease. Hence, pharmacogenetics and disease classification will be intermittently tied together as the information relating to these genetic variants become recognised.

Genomic Epidemiology – The Future of Complex Disease Genetics

Considerable progress has been made to date characterising genes that might contribute to disease in families. This linkage-based approach has provided information about multiple regions of the genome that contain determinants responsible for disease susceptibility, although to date relatively few of these genetic linkages have been converted into specific genetic variants that contribute to disease. This is likely to rely on large-scale future studies using genetic association strategies alongside genetic linkage that will allow for the characterisation of individual polymorphisms in disease populations compared to controls. There are a variety of strategies being utilised for this purpose. Perhaps the most powerful will be the utilisation of single nucleotide polymorphisms and haplotype maps defining regions of the genome demonstrating high degrees of linkage disequilibrium.[11] Linkage disequilibrium varies widely around the genome and, although certain regions such as the region around the HLA on chromosome 6 are recognised to have high levels of linkage disequilibrium, other areas show very little evidence of linkage disequilibrium even over short regions. Genetic association studies have always been highly dependent on the presence of linkage disequilibrium in that this allows polymorphisms not directly responsible for a particular phenotype to provide association data before the mechanistic polymorphism has actually been identified. As a result, many of the original HLA associations have now been refined and are some distance away from the polymorphisms that originally gave rise to the disease association. Until recently, it has not been clear how common this phenomenon was likely to be around the genome. The availability of very large numbers of SNPs has allowed such experiments now to be undertaken, and it is clear that very substantial regions of linkage disequilibrium exist on other human chromosomes besides chromosome 6.[15] In particular, an intensive study of chromosome 22 has revealed extensive patterns of haplotypic conservation that should allow rapid analysis of disease populations using relatively small numbers of SNPs. It is likely, therefore, that the future of genetic epidemiology may revolve around the characterisation of haplotypes in regions of high linkage disequilibrium as determined by studying large numbers of patients and controls. This approach will allow large numbers of patients within populations to be characterised without having to undertake excessively large sets of SNP typing. These regions can already be identified, as they are the regions of the genome with relatively low levels of recombination as identified by recombination maps. Where such regions lie within disease linkage regions they are likely to provide a rapid and efficient mechanism for identifying association in disease.

Often, this genetic information is, however, itself insufficient to provide robust information about disease pathogenesis. As a result, there is increasing interest in using a range of other methodologies to characterise gene products, both proteins and small molecules, that arise from metabolism to systematically study the pattern of these products in disease states. The availability of the whole genome sequence has provided an opportunity to generate reporter ligands for most of the gene products of interest around particular disorders that can be readily measured in blood, plasma or serum. These affinity ligands will allow the generation of protein chips and the quantification of protein markers that might be associated or prognostically implicated in particular diseases. Similarly, bio-NMR will allow the characterisation of small molecule patterns in urine and plasma that will also inform those interested in disease pathogenesis and the natural history of, particularly, metabolic and cardiovascular diseases.

Together, genetic information may best be interpreted alongside information available from these other technologies, and the patterns of expression one finds of proteins and small molecules may help to define subsets of the disease with that particular genetic susceptibility determinants. The coordinate use of genomic data broadly may help to define subsets of disease defined genetically, considerably aiding efforts to apply genetics on large epidemiological scale populations.

Conclusion

Complex trait genetics, therefore, has much to offer modern biomedicine. The characterisation of diseases based on mechanism rather than phenotype will have a profound impact, primarily on accurate disease classification and then on appropriate therapy and better prediction of natural history and appropriate management. The transformation that is likely to arise in medicine from developing a mechanistically defined understanding of disease will be enormous and the ability to recognise its importance will prove to be a crucial task for health care delivery systems and medical education. The actual delivery of genetic information that can be of use diagnostically and prognostically will increasingly rely on large patient collections characterised for sets of single nucleotide polymorphisms. The identification of regions of linkage disequilibrium will greatly facilitate this process and allow haplotypes to define disease susceptibility. This information, preferably alongside other forms of genomic information related to the effects of gene expression and their integration with the environment, will, over the next 10 years, considerably clarify our understanding of common, complex diseases.

References

1. Bell JI. The New Genetics of Clinical Practice. *B Med J* 1998; 7131:618-620.
2. Bienz M. APC: the plot thickens. *Curr Opin Genet Dev* 1999; 9:595-603.
3. Venkitaraman AR. Cancer susceptibility and the functions of BRCA1 and BRCA2. *Cell* 2002; 108:171-182.
4. Holtzman NA, Marteau TM. Will genetics revolutionize medicine? *N Engl J Med* 2000; 343:141-144.
5. Pharoah PDP, Antoniou A, Bobrow M, Zimmern RL, Easton DF, Ponder BAJ. Polygenic susceptibility to breast cancer and implications for prevention. *Nat Genet* 2002; 31:33-36.
6. Thomson G. HLA disease associations: models for the study of complex human genetic disorders. *Crit Rev Clin Lab Sci* 1995; 31:183-219.
7. McDevitt H. Closing in on type 1 diabetes. *N Engl J Med* 2001; 345:1060-1061.
8. Fajans SS, Bell GI, Polonsky KS. Molecular mechanism and clinical pathophysiology of maturity onset diabetes of the young. *N Engl J Med* 2001; 345:971-980.
9. Ryffel GU. Mutations in the human genes encoding the transcription factors of the hepatocyte nuclear factor (HNF) 1 and HNF 4 families: functional and pathological consequences. *J Mol Endocrinol* 2001; 27:11-29.
10. Stoffers DA, Zinkin NT, Stanojevic V, Clarke WI, Habener JF. Pancreatic agenesis attributable to a single nucleotide deletion in the human *IPFI* gene coding sequence. *Nat Genet* 1997; 15:106-110.
11. Chu K, Nemoz-Gaillard E, Tsai MJ. BETA 2 and pancreatic islet development. *Recent Prog Horm Res* 2001; 56:23-46.
12. Altshuler D, Hirschhorn JN, Klannemark M, Lindgren CM, Vohl MC, Nemesh J, *et al.* The common PPARgamma Pro12Ala polymorphism is associated with decreased risk of type 2 diabetes. *Nat Genet* 2000; 26:76-80.
13. Horikawa Y, Oda N, Cox NJ, Li X, Orho-Melander M, Hara M, *et al.* Genetic variation in the gene encoding calpain-10 is associated with type 2 diabetes mellitus. *Nat Genet* 2000; 26:163-175.
14. Weber WW. *Pharmacogenetics.* New York: Oxford University Press, 1997.
15. Dawson E, Abecasis GR, Bumpstead S, Chen Y, Hunt S, Beare DM, *et al.* A first-generation linkage disequilibrium map of human chromosome 22. *Nature AOP* 2002; doi:10.1038/nature00864.

11. Human Cancer Genetics

Alfred G. Knudson

Abstract

Human cancer genetics embraces both somatic and inherited mutations. The first known tumour-specific somatic aberration was the Philadelphia chromosome, which results from a translocation and activates an oncogene. This activating translocation theme pervades much of the genetics of leukaemias, lymphomas, and sarcomas. These single aberrations, in many instances, seem to be sufficient for oncogenesis, although other abnormalities often appear. No such translocation has been reported as a germline mutation. However, a few recessively inherited diseases that manifest chromosome breaks at elevated frequencies predispose to cancer.

Mendelian dominant predisposition is known for some cases of most cancers. Retinoblastoma illustrates how penetrance in hereditary cancer depends upon somatic mutation and how the same mutant gene can account for both heritable and non-hereditary cases. *RB1* is one of more than 30 cloned genes whose mutations produce heritable predisposition to cancer. Many of these are tumour suppressor genes, a few are oncogenes, and a few are DNA repair genes. The cancers most often arise after multiple somatic mutations and, at diagnosis, demonstrate chromosomal or mutational instability.

Most, perhaps even all, cancers have molecular defects that increase cell birth rate (by interfering with control of the cell cycle) and decrease cell death rate (by decreasing apoptosis in mutant cells).

Introduction

Theodor Boveri would surely be impressed, and pleased, to learn how his skeletal idea that cancer is a genetic disease of somatic cells has been fleshed out over the ensuing near century. Critical contributions to this process were the discoveries that some environmental agents known to be carcinogenic, notably ionising radiation and certain chemicals, are also mutagenic and that predisposition to cancer can be inherited in Mendelian fashion. The discovery of the Philadelphia chromosome as the first specific somatic genetic aberration associated with a particular cancer provided a capstone to this early period and has been precursor to a continuing flood of research that has provided an ever clearer idea of the origin of cancer.[1] Even the competing viral theory can be integrated into our contemporary picture of this collection of diseases.

Although heredity and environment are both causative factors for cancer, the somatic mutational concept implies that spontaneous, background mutations could be sole determinants. Furthermore, it leads to the corollary that heredity and environment may interact in carcinogenesis. Human cancers should arise then among four groups of individuals, or oncodemes[2]: persons with (1) neither hereditary predisposition nor unusual environmental exposure, (2) strong hereditary predisposition only, (3) unusual environmental exposure only, and (4)

133

both factors. We recognise that all of these groups exist, even though they are very differently distributed for different cancers. The incidences of carcinomas of the lung and cervix both reflect environmental factors, i.e., smoking and human papilloma virus, respectively, whereas, at least in the United States, the incidence of retinoblastoma reflects background mutation rates in somatic and germinal cells.

Somatic Translocations and Oncogenes

Following the discovery that the Philadelphia chromosome was formed as a result of a balanced translocation between chromosomes 9 and 22,[3] numerous leukaemias, lymphomas, and sarcomas were demonstrated to involve the same mechanism. Especially important was Burkitt's lymphoma (BL), where the 8;14 translocation that characterises a majority of cases juxtaposes the *MYC* oncogene on chromosome 8 to the immunoglobulin heavy chain gene (*IgH*) on chromosome 14, leading to pathologically elevated expression of *MYC*. [4,5] This proved to be a model for some other lymphomas and leukaemias as well. In chronic myelogenous leukeaemia (CML) the story was slightly different because the translocation occurred between part of the Abelson (*ABL*) oncogene on chromosome 9 and part of the Breakpoint Cluster Region (*BCR*) gene on chromosome 22, effectively making a new oncogene with constitutive activation of the tyrosine kinase domain of *ABL*.[6,7,8] One consequence of this activation is export of the conditional cell cycle inhibitor, p27^{kip}, to the cytoplasm, where it is ineffective;[9] another is its inhibition of apoptosis.[10] In the chronic phase of CML the karyotypes usually reveal no other cytogenetic change; the translocation appears to be both necessary and sufficient for oncogenesis, although other mutations cause progression to an acute phase of CML. In effect these cancers, and some others since, have seemed to result from a single event, a translocation.

Most of the oncogenes activated or formed by translocation are transcription factors, as exemplified by *MYC*. On one hand, *MYC* can interfere with cell cycle arrest mediated by the cyclin-dependent kinase inhibitor p21$^{WAF/CIP1}$;[11,12] on the other hand, it stimulates p53-mediated apoptosis, by an increase in protein p14^{ARF}, thus interfering with the destruction of p53 by the protein mdm2.[13,14] In BL, this latter process fails following *TP53* mutation, *MDM2* amplification, or other interference with *TP53*.[13] At least two events are required for oncogenesis in BL. For both CML and BL, oncogenesis depends upon disrupting control of the cell cycle and of apoptosis.

CML has often been reported as a delayed effect of exposure to ionising radiation (IR), and BL has a well-known increased incidence in parts of Africa, where malaria and coincidental Epstein-Barr (EB) viral infection greatly enhance the oncogenic process. The incidences of these two diseases clearly arise from at least two oncodemes; the spontaneous and the environmental only. Of interest in this context is the fact that the familial incidences of both CML and BL are extremely low. A few families with CML have been reported, but the Philadelphia chromosome itself is not inherited. These experiences appear to hold for most leukaemias, lymphomas, and sarcomas that are caused by specific balanced translocations. This phenomenon may reflect early lethality of these germline translocations; i.e., the oncogenic translocation is a dominant lethal mutation.

Acute lymphocytic leukaemia of infancy is especially interesting because of its frequent origin in a translocation that occurs during fetal life. In most cases the leukaemia cells reveal a translocation between chromosome 11 (band 11q23) and one of some 30 or more partner chromosomes.[15] These translocations interrupt the gene *MLL* that is homologous with the Trithorax gene of *Drosophila* and which is a transcription factor.[16,17,18] If the affected infant has an identical twin that shared a single monochorionic placenta, the concordance for leukaemia bearing precisely the same translocation is very high, presumably as a result of cross-circulation in fetal life[19]. This would seem to be the clearest candidate for a "single event" (translocation) cancer.

Although there are no reported cases of inheritance of a translocation that can cause non-hereditary leukaemia, lymphoma, or sarcoma, there are some recessively inherited diseases in which chromosomal breaks are abundant and can predispose to cancer, including especially leukaemia. Notable among these chromosomal breakage syndromes are Fanconi anaemia, Bloom syndrome, and Ataxia Telangiectasia (AT).

Germline Mutations and Cancer

Reports of "hereditary cancer" have long been known and include one by the French surgeon Paul Broca of multiple cases of breast cancer in his wife's family, from which Lynch has created a pedigree.[20] A famous early post-Mendelian report was that by Warthin[21] of a dominantly heritable mutation predisposing to multiple carcinomas, especially of the colon, stomach, and uterus, a condition currently called Hereditary Nonpolyposis Colon Cancer (HNPCC). Many, perhaps 50 or more, dominant predispositions are known. More than 35 responsible genes have been cloned, and another 5-10 have been mapped. The rate of identifying new genes has declined because of low incidence and/or low penetrance in the remaining conditions.

Several common threads run through the stories of these mutations. A prominent feature is that no mutant gene predisposes to all kinds of cancer. For some the tumour spectrum is very narrow or even limited to one tumour, as happens with Wilms tumour caused by germline deletion of the *WT1* gene[22]. A second common feature is that penetrance is never 100 percent; unaffected obligate mutation carriers are known for most of the hereditary cancers. Another common finding is that age at diagnosis is usually younger than for the non-hereditary form of the same cancer.

Tumour Suppressor Genes: Retinoblastoma

In all cases for which penetrance has been studied, it has proven to require at least one somatic mutation. In the first case to be clarified, retinoblastoma, somatic mutation effects alteration or loss of the second allele of the *RB1* gene, the array of mechanisms including nondisjunctional chromosome loss, somatic recombination, deletion, and intragenic mutation[23,24,25]. An expected loss of gene function ensues; i.e., the normal allele acts in effect as a tumour suppressor and its loss places the host cell on the path to cancer. For retinoblastoma this path is apparently very short and may not require any other mutations. Its pathogenesis is apparently a two-hit mechanism, which can account for its appearance even in a newborn infant. This "two-hit" tumour suppressor mechanism has been demonstrated for tumours occurring in a majority of the dominantly inherited predispositions to cancer. However, in only a few instances is the twice-hit cell malignant; in most, the resulting lesion is a hamartomatous or adenomatous precursor of cancer, and further events are required for transformation.

About 40 percent of retinoblastoma cases carry a germline mutation in *RB1*, with approximately 80 percent of them representing new germline mutations.[24] Non-hereditary retinoblastoma accounts for the other 60 percent of cases; in these tumours both alleles of the *RB1* gene are also mutant or lost, but here the events have occurred after conception. With a total incidence of about 5×10^{-5} births in the United States, the non-hereditary form occurs in 3×10^{-5}. During embryogenesis and fetal life, a small number of retinoblasts grows to an ultimate number of descendant cells in excess of 10^8.[26] This provides ample opportunity with normal mutation rates per cell division to produce one-hit clones; in fact, most persons' eyes probably contain such clones. Then one of these cells develops a second mutation or loss in 3 per 10^5 persons, a rate that is also compatible with normal somatic mutation rates, given that the once-hit clones in such subjects may be large. It is not necessary to invoke an unusual mutation rate or genomic instability. Penetrance seems to be determined solely by normal rates of somatic mutation and loss.

RB1 was the first tumour suppressor gene to be cloned.[27] Its product, pRb, is a critical regulator of entry into the G1 phase of the cell cycle. In its underphosphorylated form it sequesters an E2F transcription factor that is necessary for progression of the cycle[28,29,30,31,32]. Normal progression occurs when pRb is further phosphorylated, rendering it incapable of competing for binding of E2F. Regulated cyclic phosphorylation and dephosphorylation of pRb are key features of the cell cycle. In the absence of pRb, normal regulation does not occur and the birth rate of tumour cells increases.

Why *RB1* is a retinoblastoma gene is not known but, in fact, germline mutation in it does predispose to other tumours, especially soft tissue sarcomas and osteosarcoma[33]. These are different from the sarcomas that are caused by translocations and activation of oncogenes, such as alveolar rhabdomyosarcoma and Ewing's sarcoma. Mutation carriers may also be at increased risk of lung cancer. *RB1* is an important gene in a regulatory pathway and is mutant in the non-hereditary form of many cancers. Another gene in that pathway, *CDKN2*, whose product is p16, is mutant in many other cancers, and it is widely thought that the pRb pathway is disturbed at some point in most cancers.

Oncogenes and Hereditary Cancer

Although oncogene-activating translocations are not responsible for any hereditary cancers, a few mutant oncogenes are. Three of these genes code for tyrosine kinases that act in signal transduction: *RET* (mutant in multiple endocrine neoplasia type 2, MEN2),[34,35] *KIT* (mutant in hereditary gastrointestinal tumours, GIST)[36], and *MET* (mutant in hereditary papillary renal carcinoma, HPRC)[37]. The activated oncogene is not sufficiently potent to accomplish transformation directly. In the case of HPRC, the tumours are trisomic for the chromosome (number 7) that carries the *MET* oncogene; two of these are mutant and one is normal.[38] Evidently, two mutants exceed some threshold of transformation. Since the karyotypes may be otherwise normal, these tumours seem to result from two hits that involve the same gene, despite the presence of one normal allele. A similar phenomenon occurs in MEN2, where the tumours may also be trisomic, in some cases losing the wild type *RET* allele.[39]

The Li-Fraumeni Syndrome and *TP53*

The Li-Fraumeni syndrome (LFS) was discovered through the study of familial rhabdomyosarcoma in children[40] and was later shown, in many cases, to be caused by mutation in *TP53*[41], a gene discovered through the study of a cellular protein p53, which associated with transforming proteins of certain DNA tumour viruses, such as the large T antigen of Simian Virus 40 (SV40).[42,43] The chief clinical manifestation of LFS is breast cancer, which affects many female mutant carriers by the age of 50 years. Like hereditary retinoblastoma, LFS predisposes strongly to soft tissue sarcomas and osteosarcoma, so it is not surprising that the non-hereditary forms of these two categories of sarcoma are also often defective for *TP53*. What is surprising is that many non-hereditary carcinomas, including those of the colon and pancreas, have very high incidences of homozygous mutation/loss of *TP53*, but are not featured in LFS.

TP53 mediates conditional responses to DNA damage, as inflicted by IR and certain chemicals.[44] The level of p53 protein is normally low, but is increased following exposure to IR. The responses may be cell cycle arrest and subsequent repair of the lesions or induction of apoptosis.[45,46,47] In the presence of mutant *TP53*, one response is failure of apoptosis, survival of abnormal chromosomes, and ensuing chromosomal instability (CIN), which is a frequent development because *TP53* is the most frequently mutated gene in cancer. Just as defects in the pRb pathway lead to an increased birth rate of cancer cells, loss or abnormality of p53 causes a decrease in their death rate.

Most carcinomas are defective in the regulation of both the cell cycle and apoptosis. Thus, in familial adenomatous polyposis (FAP), a polyp results from loss or mutation of both alleles of the adenomatous polyposis coli (*APC*) gene and is subsequently transformed into a carcinoma following *TP53* mutation/loss.[48,49,50,51] The defect in *APC* eliminates its ability to degrade β-catenin, which can transmit signals for mitosis to the nucleus, one result being stimulation of the *MYC* oncogene, which can, in turn, abrogate control of the cell cycle by pRb. This pattern of defect in one gene that is important for regulation of signal transduction and in another for apoptosis is a recurring theme in carcinomas. Such tumours have sustained at least four "hits", i.e., two in a cell cycle regulatory gene and two in a regulator of apoptosis, most often *TP53*. Furthermore, loss of APC protein leads to failure of chromosomes to connect normally with kinetochores, so leading to aneuploidy and chromosomal aberrations.[52,53,54] This observation that mutant *APC* can lead to chromosomal abnormalities may explain the genetic instability previously observed in adenomatous polyps.[55,56] The fact that gross karyotypic aberrations are not typically observed may be the result of DNA repair in some cells and of apoptosis induced by *TP53* in others; when *TP53* becomes mutated at the transition to carcinoma, aberrations would appear in abundance (CIN). Following loss of *TP53* and the onset of CIN, other mutations, losses, and translocations can affect genes that are responsible for invasion and metastasis.

HNPCC, Breast Cancer, and DNA Repair

One of the most common forms of hereditary cancer is Hereditary Non-Polyposis Colon Cancer (HNPCC), which, as noted previously, also predisposes to several other cancers. Its incidence is greater than one per thousand persons and new mutants represent a small fraction of all cases. Several different genes are responsible, but most cases are mutant heterozygotes for *MSH2* or *MLH1*, human homologues of DNA mismatch repair genes known in bacteria and yeast.[57,58,59,60] Heterozygous target cells first acquire a mutation in the second allele of the gene, thereby producing *microsatellite instability* (MIN).[61] Mutation rates in homozygous cells are of the order of 10^3 times normal rates.[62] Particularly susceptible are poly A stretches of DNA, often 8-10 in length, which contract or lengthen in successive cell cycles. Digested tumour DNAs reveal new microsatellite bands. In the colon the TGFβ receptor 2 gene (TGFBR2), a tumour suppressor, is the prime target in HNPCC. Cancer incidence is high because the high mutation rates have a multiplicative effect that compensates for the extra mutation (in the second allele of *MSH2* or *MLH1*) that occurs at a normal rate. The tumours that result usually have normal, or nearly normal, karyotypes, i.e., they do not show CIN. It now appears that no cancer develops both MIN and CIN; they are two different kinds of genomic instability.[63]

The discovery of a DNA repair defect in heterozygotes was a surprise because previous such defects were recessively inherited. Presumably, selection has been operating against homozygotes rather than heterozygotes, hence the high heterozygote frequencies and low fraction of new mutations in comparison to previously discussed dominantly inherited cancers. As expected, profound founder effects are also noted in different populations. Two other examples of this phenomenon in populations have emerged from the study of hereditary breast cancer, where two genes, *BRCA1* and *BRCA2,* display these same epidemiological features. Again, tumour specificity is not as complete as suggested by disease or gene names. These two genes also operate in DNA repair, in protein complexes that mediate responses to DNA damage, notably DNA double strand breaks (DSBs), and in conjunction with the Ataxia Telangiectasia gene (*ATM*).[64] With loss of *BRCA1* or *BRCA2,* cells do not repair DSBs by homologous recombination, but rather by non-homologous end-joining. [65]

Conclusions

For the geneticist the collection of diseases known as cancer is unique in that both somatic and germline mutations are important in their origin. Some cancers belong primarily to the somatic category and involve one or, at most, very few somatic genetic events, primarily balanced translocations, in their pathogenesis. Other cancers, including most, if not all, carcinomas, arise following selectable mutations in more than one, but apparently only a few, genes and occur in both hereditary and non-hereditary form. Penetrance in the hereditary forms requires somatic mutations, one of which is most frequently in the second allele of the gene whose germline mutation imposed predisposition to cancer. In most cancers the mutations, whether somatic only or both somatic and germinal, provide a growth advantage by disturbing regulation of the cell cycle or a survival advantage by interfering with apoptotic mechanisms.

References

1   Nowell, P.C. and Hungerford, D.A. A minute chromosome in human chronic granulocytic leukemia. *Science* 132:1497, 1960.

2   Knudson, A.G. Hereditary cancer, oncogenes, and antioncogenes. *Cancer Res.* 45:1437-1443, 1985.

3   Rowley, J.D. A new consistent chromosomal abnormality in chronic myelogenous leukaemia identified by quinacrine fluorescence and Giemsa staining. *Nature* 243:290-293, 1973.

4   Dalla-Favera, R., Bregni, M., Erikson, J., Patterson, D., Gallo, R.C., and Croce, C.M. Human c-myc onc gene is located on the region of chromosome 8 that is translocated in Burkitt lymphoma cells. *Proc. Natl. Acad. Sci. USA* 79:7824-7827, 1982.

5   Taub, R., Kirsch, I., Morton, C., Lenoir, G., Swan, D., Tronick, S., Aaronson, S., and Leder, P. Translocation of the c-myc gene into the immunoglobulin heavy chain locus in human Burkitt lymphoma and murine plasmacytoma cells. *Proc. Natl. Acad. Sci. USA* 79:7837-7841, 1982.

6   Konopka, J.B., Watanabe, S.M., Singer, J.W., Collins, S.J., and Witte, O.N. Cell lines and clinical isolates derived from Ph1-positive chronic myelogenous leukemia patients express c-abl proteins with a common structural alteration. *Proc. Natl. Acad. Sci. USA* 82:1810-1814, 1985.

7   Shtivelman, E., Lifshitz, B., Gale, R.P., and Canaani, E. Fused transcript of abl and bcr genes in chronic myelogenous leukaemia. *Nature* 315:550-554, 1985.

8   Stam, K., Heisterkamp, N., Grosveld, G., de Klein, A., Verma, R.S., Coleman, M., Dosik, H., and Groffen, J. Evidence of a new chimeric bcr/c-abl mRNA in patients with chronic myelocytic leukemia and the Philadelphia chromosome. *N. Engl. J. Med.* 313:1429-1433, 1985.

9   Jiang, Y., Zhao, R.C., and Verfaillie, C.M. Abnormal integrin-mediated regulation of chronic myelogenous leukemia CD34+ cell proliferation: BCR/ABL up-regulates the cyclin-dependent kinase inhibitor, p27^{Kip}, which is relocated to the cell cytoplasm and incapable of regulating cdk2 activity. *Proc. Natl. Acad. Sci. USA* 97:10538-10543, 2000.

10 Horita, M., Andreu, E.J., Benito, A., Arbona, C., Sanz, C., Benet, I., Prosper, F., and Fernandez-Luna, J.L. Blockade of the Bcr-Abl kinase activity induces apoptosis of chronic myelogenous leukemia cells by suppressing signal transducer and activator of transcription stat5-dependent expression of Bcl-xL. *J. Exp. Med.* 191:977-984, 2000.

11 Coller, H.A., Grandori, C., Tamayo, P., Colbert, T., Lander, E.S., Eisenman, R.N., and Golub, T.R. Expression analysis with oligonucleotide microarrays reveals that MYC regulates genes involved in growth, cell cycle, signaling, and adhesion. *Proc. Natl. Acad. Sci. USA* 97:3260-3265, 2000.

12 Gartel, A.L., Ye, X., Goufman, E., Shianov, P., Hay, N., Najmabadi, F., and Tyner, A.L. Myc represses the p21(WAF1/CIP1) promoter and interacts with Sp1/Sp3. *Proc. Natl. Acad. Sci. USA* 98:4510-4515, 2001.

13 Goetz, A.W., van Der Kuip, H., Maya, R., Oren, M., and Aulitzky, W.E. Requirement for Mdm2 in the survival effects of Bcr-Abl and interleukin 3 in hematopoietic cells. *Cancer Res.* 61:7635-7641, 2001.

14 Vogelstein, B., Lane, D., and Levine, A.J. Surfing the p53 network. *Nature* 408:307-310, 2000.

15 Rowley, J.D. The critical role of chromosome translocations in human leukemias. *Annu. Rev. Genet.* 32:495-519, 1998.

16 Djabali, M., Selleri, L., Parry, P., Bower, M., Young, B.D., and Evans, G.A. A trithorax-like gene is interrupted by chromosome 11q23 translocations in acute leukaemias [published erratum appears in *Nat. Genet.* 1993 Aug;4(4):431]. *Nat. Genet.* 2:113-118, 1992.

17 Gu, Y., Nakamura, T., Alder, H., Prasad, R., Canaani, O., Cimino, G., Croce, C.M., and Canaani, E. The t(4;11) chromosome translocation of human acute leukemias fuses the ALL-1 gene, related to *Drosophila trithorax*, to the *AF-4* gene. *Cell* 71:701-708, 1992.

18 Tkachuk, D.C., Kohler, S., and Cleary, M.L. Involvement of a homolog of *Drosophila trithorax* by 11q23 chromosomal translocations in acute leukemias. *Cell* 71:691-700, 1992.

19 Greaves, M. Molecular genetics, natural history and the demise of childhood leukaemia. *Eur. J. Cancer* 35:1941-1953, 1999.

20 Lynch, H.T. Introduction to cancer genetics. In: *Cancer Genet.*, H.T. Lynch, Ed. 1976, Charles C. Thomas Pub., Springfield, IL. p. 3-31.

21 Warthin, A.S. Heredity with reference to carcinoma. *Arch. Intern. Med.* 12:546-555, 1913.

22 Call, K.M., Glaser, T., Ito, C.Y., Buckler, A.J., Pelletier, J., Haber, D.A., Rose, E.A., Kral, A., Yeger, H., Lewis, W.H., et al. Isolation and characterization of a zinc finger polypeptide gene at the human chromosome 11 Wilms' tumor locus. *Cell* 60:509-520, 1990.

23 Cavenee, W.K., Dryja, T.P., Phillips, R.A., Benedict, W.F., Godbout, R., Gallie, B.L., Murphree, A.L., Strong, L.C., and White, R.L. Expression of recessive alleles by chromosomal mechanisms in retinoblastoma. *Nature* 305:779-784, 1983.

24 Knudson, A.G. Mutation and cancer: statistical study of retinoblastoma. *Proc. Natl. Acad. Sci. USA* 68:820-823, 1971.

25 Knudson, A.G. Retinoblastoma: a prototypic hereditary neoplasm. *Semin. Oncol.* 5:57-60, 1978.

26 Hethcote, H.W. and Knudson, A.G. Model for the incidence of embryonal cancers: application to retinoblastoma. *Proc. Natl. Acad. Sci. USA* 75:2453-2457, 1978.

27 Friend, S.H., Bernards, R., Rogelj, S., Weinberg, R.A., Rapaport, J.M., Albert, D.M., and Dryja, T.P. A human DNA segment with properties of the gene that predisposes to retinoblastoma and osteosarcoma. *Nature* 323:643-646, 1986.

28 Bagchi, S., Weinmann, R., and Raychaudhuri, P. The retinoblastoma protein copurifies with E2F-I, an E1A-regulated inhibitor of the transcription factor E2F. *Cell* 65:1063-1072, 1991.

29 Chellappan, S.P., Hiebert, S., Mudryj, M., Horowitz, J.M., and Nevins, J.R. The E2F transcription factor is a cellular target for the RB protein. *Cell* 65:1053-1061, 1991.

30 Helin, K., Lees, J.A., Vidal, M., Dyson, N., Harlow, E., and Fattaey, A. A cDNA encoding a pRB-binding protein with properties of the transcription factor E2F. *Cell* 70:337-350, 1992.

31 Kaelin, W.G., Krek, W., Sellers, W.R., Decaprio, J.A., Ajchenbaum, F., Fuchs, C.S., Chittenden, T., Li, Y., Farnham, P.J., Blanar, M.A., Livingston, D.M., and Flemington, E.K. Expression cloning of a cDNA encoding a retinoblastoma-binding protein with E2F-like properties. *Cell* 70:351-364, 1992.

32 Shirodkar, S., Ewen, M., DeCaprio, J.A., Morgan, J., Livingston, D.M., and Chittenden, T. The transcription factor E2F interacts with the retinoblastoma product and a p107-cyclin A complex in a cell cycle-regulated manner. *Cell* 68:157-166, 1992.

33 Draper, G.J., Sanders, B.M., and Kingston, J.E. Second primary neoplasms in patients with retinoblastoma. *Br. J. Cancer* 53:661-71, 1986.

34 Donis-Keller, H., Dou, S., Chi, D., Carlson, K.M., Toshima, K., Lairmore, T.C., Howe, J.R., Moley, J.F., Goodfellow, P., and Wells, S.A. Mutations in the RET proto-oncogene are associated with MEN 2A and FMTC. *Hum. Mol. Genet.* 2:851-856, 1993.

35 Mulligan, L.M., Kwok, J.B., Healey, C.S., Elsdon, M.J., Eng, C., Gardner, E., Love, D.R., Mole, S.E., Moore, J.K., Papi, L., Ponder, M.A., Telenlus, H., Tunnacliffe, A., and Ponder, B.A.J. Germ-line mutations of the RET proto-oncogene in multiple endocrine neoplasia type 2A. *Nature* 363:458-460, 1993.

36 Nishida, T., Hirota, S., Taniguchi, M., Hashimoto, K., Isozaki, K., Nakamura, H., Kanakura, Y., Tanaka, T., Takabayashi, A., Matsuda, H., and Kitamura, Y. Familial gastrointestinal stromal tumours with germline mutation of the KIT gene. *Nat. Genet.* 19:323-324, 1998.

37 Schmidt, L., Duh, F.M., Chen, F., Kishida, T., Glenn, G., Choyke, P., Scherer, S.W., Zhuang, Z.P., Lubensky, I., Dean, M., Allikmets, R., Chidambaram, A., Bergerheim, U.R., Feltis, J.T., Casadevall, C., Zamarron, A., Bernues, M., Richard, S., Lips, C.J.M., Walther, M.M., Tsui, L.C., Geil, L., Orcutt, M.L., Stackhouse, T., Lipan, J., Slife, L., Brauch, H., Decker, J., Niehans, G., Hughson, M.D., Moch, H., Storkel, S., Lerman, M.I., Linehan, W.M., and Zbar, B. Germline and somatic mutations in the tyrosine kinase domain of the MET proto-oncogene in papillary renal carcinomas. *Nat. Genet.* 16:68-73, 1997.

38 Zhuang, Z., Park, W.S., Pack, S., Schmidt, L., Vortmeyer, A.O., Pak, E., Pham, T., Weil, R.J., Candidus, S., Lubensky, I.A., Linehan, W.M., Zbar, B., and Weirich, G. Trisomy 7-harboring non-random duplication of the mutant MET allele in hereditary papillary renal carcinomas. *Nat. Genet.* 20:66-69, 1998.

39 Huang, S.C., Koch, C.A., Vortmeyer, A.O., Pack, S.D., Lichtenauer, U.D., Mannan, P., Lubensky, I.A., Chrousos, G.P., Gagel, R.F., Pacak, K., and Zhuang, Z. Duplication of the mutant RET allele in trisomy 10 or loss of the wild-type

allele in multiple endocrine neoplasia type 2-associated pheochromocytomas. *Cancer Res.* 60:6223-6226, 2000.

40 Li, F.P. and Fraumeni, J.F. Soft-tissue sarcomas, breast cancer, and other neoplasms. A familial syndrome? *Ann. Intern. Med.* 71:747-752, 1969.

41 Malkin, D., Li, F.P., Strong, L.C., Fraumeni, J.F., Nelson, C.E., Kim, D.H., Kassel, J., Gryka, M.A., Bischoff, F.Z., Tainsky, M.A., and Friend, S.H. Germ line p53 mutations in a familial syndrome of breast cancer, sarcomas, and other neoplasms. *Science* 250:1233-1238, 1990.

42 Lane, D.P. and Crawford, L.V. T antigen is bound to a host protein in SV40-transformed cells. *Nature* 278:261-263, 1979.

43 Linzer, D.I. and Levine, A.J. Characterization of a 54K dalton cellular SV40 tumor antigen present in SV40-transformed cells and uninfected embryonal carcinoma cells. *Cell* 17:43-52, 1979.

44 Kastan, M.B., Onyekwere, O., Sidransky, D., Vogelstein, B., and Craig, R.W. Participation of p53 protein in the cellular response to DNA damage. *Cancer Res.* 51:6304-6311, 1991.

45 Matsuoka, S., Huang, M., and Elledge, S.J. Linkage of ATM to cell cycle regulation by the Chk2 protein kinase. *Science* 282:1893-1897, 1998.

46 Yin, Y., Tainsky, M.A., Bischoff, F.Z., Strong, L.C., and Wahl, G.M. Wild-type p53 restores cell cycle control and inhibits gene amplification in cells with mutant p53 alleles. *Cell* 70:937-48, 1992.

47 Yonish-Rouach, E., Resnitzky, D., Lotem, J., Sachs, L., Kimchi, A., and Oren, M. Wild-type p53 induces apoptosis of myeloid leukaemic cells that is inhibited by interleukin-6. *Nature* 352:345-347, 1991.

48 Fearon, E.R. and Vogelstein, B. A genetic model for colorectal tumorigenesis. *Cell* 61:759-767, 1990.

49 Groden, J., Thliveris, A., Samowitz, W., Carlson, M., Gelbert, L., Albertsen, H., Joslyn, G., Stevens, J., Spirio, L., Robertson, M., Sargeant, L., Krapcho, K., Wolff, E., Burt, R., Hughes, J.P., Warrington, J., Mcpherson, J., Wasmuth, J., Lepaslier, D., Abderrahim, H., Cohen, D., Leppert, M., and White, R. Identification and characterization of the familial adenomatous polyposis coli gene. *Cell* 66:589-600, 1991.

50 Kikuchi-Yanoshita, R., Konishi, M., Ito, S., Seki, M., Tanaka, K., Maeda, Y., Iino, H., Fukayama, M., Koike, M., and Mori, T. Genetic changes of both p53 alleles associated with the conversion from colorectal adenoma to early carcinoma in familial adenomatous polyposis and non-familial adenomatous polyposis patients. *Cancer Res.* 52:3965-3971, 1992.

51 Nishisho, I., Nakamura, Y., Miyoshi, Y., Miki, Y., Ando, H., Horii, A., Koyama, K., Utsunomiya, J., Baba, S., and Hedge, P. Mutations of chromosome 5q21 genes in FAP and colorectal cancer patients. *Science* 253:665-669, 1991.

52 Fodde, R., Kuipers, J., Rosenberg, C., Smits, R., Kielman, M., Gaspar, C., van Es, J.H., Breukel, C., Wiegant, J., Giles, R.H., and Clevers, H. Mutations in the APC tumour suppressor gene cause chromosomal instability. *Nat. Cell. Biol.* 3:433-438, 2001.

53 Kaplan, K.B., Burds, A.A., Swedlow, J.R., Bekir, S.S., Sorger, P.K., and Nathke, I.S. A role for the Adenomatous Polyposis Coli protein in chromosome segregation. *Nat. Cell. Biol.* 3:429-432, 2001.

54 Pellman, D. A CINtillating new job for the APC tumor suppressor. *Science* 291:2555-2556, 2001.

55 Shih, I.M., Zhou, W., Goodman, S.N., Lengauer, C., Kinzler, K.W., and Vogelstein, B. Evidence that genetic instability occurs at an early stage of colorectal tumorigenesis. *Cancer Res.* 61:818-822, 2001.

56 Stoler, D.L., Chen, N., Basik, M., Kahlenberg, M.S., Rodriguez-Bigas, M.A., Petrelli, N.J., and Anderson, G.R. The onset and extent of genomic instability in sporadic colorectal tumor progression. *Proc. Natl. Acad. Sci. USA* 96:15121-15126, 1999.

57 Bronner, C.E., Baker, S.M., Morrison, P.T., Warren, G., Smith, L.G., Lescoe, M.K., Kane, M., Earabino, C., Lipford, J., Lindblom, A., Tannergard, P., Bollag, R.J., Godwin, A.R., Ward, D.C., Nordenskjøld, M., Fishel, R., Kolodner, R., and Liskay, M. Mutation in the DNA mismatch repair gene homologue hMLH1 is associated with hereditary non-polyposis colon cancer. *Nature* 368:258-261, 1994.

58 Fishel, R., Lescoe, M.K., Rao, M.R., Copeland, N.G., Jenkins, N.A., Garber, J., Kane, M., and Kolodner, R. The human mutator gene homolog MSH2 and its association with hereditary nonpolyposis colon cancer. *Cell* 75:1027-1038, 1993.

59 Leach, F.S., Nicolaides, N.C., Papadopoulos, N., Liu, B., Jen, J., Parsons, R., Peltomäki, P., Sistonen, P., Aaltonen, L.A., Nyström-Lahti, M., Guan, X.-Y., Zhang, J., Meltzer, P.S., Yu, J.-W., Kao, F.-T., Chen, D.J., Cerosaletti, K.M., Fournier, R.E.K., Todd, S., Lewis, T., Leach, R.L., Naylor, S.L., Weissenbach, J., Mecklin, J.-P., Järvinen, H., Petersen, G.M., Hamilton, S.R., Green, J., Jass, J., Watson, P., Lynch, H.T., Trent, J.M., de la Chapelle, A., Kinzler, K.W., and Vogelstein, B. Mutations of a mutS homolog in hereditary nonpolyposis colorectal cancer. *Cell* 75:1215-1225, 1993.

60 Papadopoulos, N., Nicolaides, N.C., Wei, Y.F., Ruben, S.M., Carter, K.C., Rosen, C.A., Haseltine, W.A., Fleischmann, R.D., Fraser, C.M., Adams, M.D., Venter, J.C., Hamilton, S.R., Petersen, G.M., Watson, P., Lynch, H.T., Peltomaki, P., Mecklin, J.P., Delachapelle, A., Kinzler, K.W., and Vogelstein, B. Mutation of a mutL homolog in hereditary colon cancer. *Science* 263:1625-1629, 1994.

61 Ionov, Y., Peinado, M.A., Malkhosyan, S., Shibata, D., and Perucho, M. Ubiquitous somatic mutations in simple repeated sequences reveal a new mechanism for colonic carcinogenesis. *Nature* 363:558-561, 1993.

62 Bhattacharyya, N.P., Skandalis, A., Ganesh, A., Groden, J., and Meuth, M. Mutator phenotypes in human colorectal carcinoma cell lines. *Proc. Natl. Acad. Sci. USA* 91:6319-6323, 1994.

63 Bardelli, A., Cahill, D.P., Lederer, G., Speicher, M.R., Kinzler, K.W., Vogelstein, B., and Lengauer, C. Carcinogen-specific induction of genetic instability. *Proc. Natl. Acad. Sci. USA* 98:5770-5775, 2001.

64 Khanna, K.K. and Jackson, S.P. DNA double-strand breaks: signaling, repair and the cancer connection. *Nat. Genet.* 27:247-254, 2001.

65 Wang, H., Zeng, Z.C., Bui, T.A., DiBiase, S.J., Qin, W., Xia, F., Powell, S.N., and Iliakis, G. Nonhomologous end-joining of ionizing radiation-induced DNA double-stranded breaks in human tumor cells deficient in BRCA1 or BRCA2. *Cancer Res.* 61:270-277, 2001.

12. Genetics and the Future of Medicine

D. J. Weatherall

Introduction

The announcement of the partial completion of the Human Genome Project was accompanied by some remarkable predictions about the benefits for human health that would result from this extraordinary achievement. The media and even some of the scientists involved predicted that it would completely transform medical practice over the next 20 years and would provide a way of preventing or curing most of our intractable diseases. It is not surprising that many doctors were surprised to hear this and not a little sceptical about these claims. After all, their exposure to genetics, even the younger members of the profession, was limited and, apart from a few rare diseases, most of which they have never seen, genetics appeared to play very little role in their day to day clinical practice.

The widely misunderstood reason for the expectations about the benefits of genomics for human health is based on a much broader view of "genetics" than has been taken in the past by the medical profession, or most biologists for that matter. All living organisms, whether in health or disease, are what they are by virtue of their genetic make-up, their environment, and the long and constantly changing history of the cultures in which they are raised. Many of the consequences of these complex interactions can, it is held, ultimately be explained at a biochemical level. Hence, since all biochemical reactions are regulated by the genome, an understanding of gene action and its variation should offer us a much better appreciation of the basic mechanisms of life, both in health and in disease.

This broader definition of the importance of genetics in medicine encompasses the conventional view that some diseases result from single defective genes or from increased susceptibility to environmental agents due to variation in the activity of a number of different genes. But it goes much further than this. It recognises that disease may also result from damage to the genome acquired by exposure to environmental or endogenous factors over our lifetimes and that this type of process may also be part of the complex pathology of cancer and ageing. It also stresses the importance of our evolutionary histories, suggesting that a genetic make-up which was selected for the completely different environments of our hunter-gatherer forebears may not have had time to adapt to the completely different conditions in which we find ourselves in today. And it extends our thinking about the importance of genetics beyond the human genome to those of the innumerable pathogens that we have been unable to control and which still decimate large populations of the world.

Before considering the extent to which our current hopes for genomics and better health are likely to come to fruition, and if so when, it is important to consider what the major medical problems are likely to be in the 21st century.

Disease in the 21st Century

The 20th century saw a remarkable improvement in the health of many countries. Due to improvements in nutrition and hygiene and the development of powerful vaccines and antibiotics, many common infectious diseases were largely controlled. The life expectancy of populations increased dramatically and in richer countries infectious killers were replaced by the intractable diseases of middle and old age, notably vascular disease and cancer.

As many of the poorer countries of the world have gone through the demographic transition following improvements in living conditions and in their economies, their pattern of disease has changed in the same way. Indeed, globally, ischaemic heart disease is now the commonest cause of death while, by the year 2020, it is estimated that bipolar affective disorders may be the commonest cause of chronic ill health (WHO, 2000). In addition, road accidents, tobacco-related disease which is increasing rapidly in the developing countries, and other problems related to the growing stresses of modern life will become major health problems.

However, despite these remarkable improvements in health, there is still what the World Health Organization (WHO) refers to as the "unfinished agenda". Many populations still live in dire poverty and have extremely high infant mortality and relatively short life expectancies. The three major infections among the many which have not been controlled, AIDS, malaria and tuberculosis, are increasingly important killers. Because of globalisation, no country is now protected from these diseases, and AIDS and tuberculosis are presenting an increasing problem in the richer countries. Furthermore, new and drug-resistant organisms are appearing all the time, and as this chapter is being prepared, the long-feared vista of biological warfare has become a reality.

The Spectrum of Genetic Medicine for the Future

The overall spectrum of the application of genomics for medical research and health care is summarised in Table 12.1. In short, it is clear that the tools of genomics, particularly when they are developed to provide a more functional and integrative picture of how our 30,000 genes function and interact with one another, will have application right across the field of medical practice.

In the sections that follow, I shall attempt to outline the more important of these applications and hazard a guess about how long it may take before they are of value in the clinic.

Monogenic Disease

Although most monogenic diseases are rare, because there are some 5,000 of them, collectively they form an important part of paediatric practice. And in populations in which these conditions have reached very high frequencies, particularly the inherited disorders of haemoglobin including the sickle cell disorders and the thalassaemias, they are producing an increasing health burden, a major problem since most of these high frequency regions involve countries of the developing world. Between 1981 and 2000, 1,112 disease genes were discovered, as well as 94 disease-related genes in various forms of cancer (see

Table 12.1: Spectrum of potential applications of genomics to medical practice

Monogenic disease

- Mechanisms. Heterogeneity
- Diagnosis. Counselling. Control
- ? Therapy

Developmental disorders

- Monogenic. Chromosomal abnormalities
- Multifactorial

Common multifactorial disease

- Susceptibility genes. Disease mechanisms. New drugs
- ? Preventative medicine

Somatic mutation

- Cancer. Diagnosis. Prognosis. Therapy
- Ageing

Pathogen and vector genomics

- Vaccines and therapy
- Reduce transmission rates

Broader issues of human biology

- Development
- Evolution
- Neurobiology

later section). At least in a few cases it has been possible to start to explain the relationship between the disease phenotype and the underlying mutation.

As knowledge of the monogenic disease has accumulated, it has become increasingly apparent that they show remarkable clinical heterogeneity, even within families with the same mutation. For the most part, the reasons are not yet clear. However, in the thalassaemia field it has been found that it reflects the action of layer upon layer of modifier genes, which may reduce or increase the severity of the action of the mutant gene or which may act at a distance and modify the complications of the disease. Environmental factors also modify the phenotype of the thalassaemias, as do local selective factors which vary from population to population. Thus, even what at first sight appeared to be a simple monogenic condition may have a profoundly different phenotype depending on the action of genetic modifiers and environmental factors.

Despite these complexities, this new information has already been used widely for carrier detection, population screening and prenatal diagnosis, and has revolutionised this aspect of medical practice. For example, the development of population screening programmes together with prenatal diagnosis has led to a dramatic reduction in the frequency of births of patients with β thalassaemia in some of the Mediterranean islands and mainland populations.

Many problems remain, however. Before really accurate genetic counselling can be applied widely, more will have to be learnt about the reasons for the phenotypic heterogeneity of many monogenic diseases. Furthermore, screening using DNA technology is relatively expensive and, because of the large number of mutations that are involved in many monogenic diseases, many hundreds in some cases, is still time consuming. The great success of the thalassaemia field has depended to no small degree on the development of cheaper, non-DNA based screening methods, an objective which should be sought for all monogenic diseases.

In truth, DNA technology has done very little towards the management of monogenic disease. Somatic-cell gene therapy has proved much more difficult than was originally hoped, although there have been a few limited successes. It seems likely that the technological problems will be solved, at least in the case of monogenic diseases due to mutations in housekeeping genes, that is, genes which are expressed in most cells at a very low level. The correction of genetic defects which require high level, tissue-specific expression may take a great deal longer.

In short, sufficient progress has been made in the application of genomics to the study of monogenic disease to suggest that this new technology is already an established part of clinical practice, at least in developed countries, and that slow but steady progress will be made towards the correction of at least a few of these disorders. It is too early to estimate the cost of somatic-cell gene therapy, although it seems likely that it will be very expensive and restricted to the more advanced countries for the foreseeable future.

Communicable Disease

So far, information obtained from the different pathogen genome projects has had very limited clinical application. A few diagnostic agents have been developed which are turning out to have particular value for the identification of organisms which are difficult to grow in culture, and some progress has been made towards defining genes in human populations which convey resistance to particular pathogens or to the treatment of communicable disease.

There are, however, grounds for optimism. Two examples of the way in which the pathogen genome is being exploited point towards the way the field may develop in the future. Two genes have been identified in the genome of *Plasmodium falciparum*, the organism which causes the severe form of malaria, that encode enzymes in two key biochemical pathways of the parasite. Inhibitors of these enzymes have been developed, both of which reduce the growth of the parasite in mice and neither of which have proved toxic, suggesting that they may form the basis for new classes of anti-malarial drugs. As an approach to using genomics to search for vaccine candidates, the entire genome sequence of a virulent strain of *Neisseria meningitidis* was searched, and over 500 cell-surface-expressed or secreted proteins were identified. The corresponding DNA sequences were cloned in bacteria and over half were expressed successfully and used to immunise mice. Following this large screening procedure two highly conserved vaccine candidates emerged.

These are early days, but, considering the increasing problem of infectious disease, it is reasonable to hope that the Pathogen Genome Project will provide similar approaches to chemotherapy and vaccine production although there will be a long gap between the identification of potential candidates and the development of agents that are valuable in the clinic. Promising results are also being obtained in efforts to modify the genomes of vectors, for example mosquitoes, to reduce their ability to transmit disease.

Cancer

The spectacular progress that has already been made in the application of molecular and cell biology to the study of cancer has reflected the amalgamation of knowledge obtained from classical epidemiology, cytogenetics, cell biology, and tumour virology. The recognition that many cancers result from the acquisition of mutations of cellular oncogenes, which may be the result of life-long exposure to external carcinogens or to the powerful oxygens, which are being produced continually as part of normal body metabolism, or their abnormal activation or deregulation as the result of specific chromosomal abnormalities, has provided completely new insights into the genesis of neoplastic transformation. These observations are leading to some fundamental rethinking about the prevention and management of cancer. It should, for example, be possible to classify particular tumours according to the expression of different sets of oncogenes, an observation which has already been confirmed for certain cancers of the blood or breast. The hope is that cancer therapy will change from the hit-and-miss approach of destroying both malignant and healthy cells to treatment that is more functionally direct.

Many problems remain, however. Increasing experience of the study of cancer at the molecular level highlights the remarkably heterogeneous pathways to the development of cancer. It is starting to look as though cancer therapy may have to be custom made for many different tumours, even those involving the same organ and the same cellular morphology. And cancer cells have, in common with micro-organisms, the ability rapidly to mutate, leading to resistance to chemotherapy. The first great success story of the molecular approach to cancer treatment is a good example. It has been known for years that chronic myeloid leukaemia is associated with a chromosomal translocation that produces a fusion gene which encodes for a novel tyrosine kinase. Remarkably, it was possible to manufacture a drug directed specifically at this kinase, which had the effect of producing remission in this common form of leukaemia. However, within a short time it became apparent that cell populations were emerging that were resistant to this new agent due to the production of a variant kinase. There is no reason to believe that drug resistance will be any less common when agents are directed at variants of oncogene products than it is in our current blunderbuss chemotherapy approach to the management of cancer.

Again, therefore, the fruits of genomics in the cancer field may take a long time to reach the clinic and even longer before we can be sure that these new approaches to treatment are better than those that we have at the present time.

Complex Multifactorial Diseases

The major killers of affluent societies, and indeed those that are passing through the demographic transition, are heart disease, stroke and diabetes. But in terms of a drain on health resources, many other chronic intractable diseases, particularly the important psychiatric disorders, rheumatism, autoimmune disease, asthma and others, are an increasing drain on health-care provision. With the major increase in the size of the ageing population, the dementias are also becoming an increasingly important burden. It seems likely that all these conditions reflect interactions between the environment, a varying degree of genetic susceptibility and, almost certainly, the complex and still ill-understood pathology of ageing. It is for the control and management of these conditions that the greatest claims for the benefits of genomics have been made. It is believed that by carrying out large-scale linkage or association studies using markers such as single nucleotide polymorphisms (SNPs) or haplotypes thereof, that is, inherited blocks of these polymorphisms, it may be possible to characterise some of the genes that are involved in susceptibility or resistance to these conditions. If this is possible, analysing the action of these genes should lead to a better understanding of the underlying pathophysiology of these conditions, information which should lead to more focused forms of treatment and, ultimately, to the identification of high risk groups for which public health measures may be focused. Even promises of early diagnosis with genetic correction for susceptibility to some of these conditions have been put on the agenda.

So far, this field has met with relatively little success and has thrown up many false leads. A few loci have been defined, but an enormous amount of work has come to nothing. This should not surprise us, however. Most of these conditions have extremely complex phenotypes and are probably quite heterogeneous in their basic pathology. They are all disorders of middle or old age, and hence their pathology may well be complicated by the complex pathophysiology of ageing itself. There are formidable technical problems in carrying out association and linkage studies involving these multigenic systems, including population heterogeneity and difficulties in determining the relative importance of the action of a number of genes, some of which probably have a very small effect.

In some cases disorders of this type occur at an earlier age and appear to be determined by a single gene, for example, some forms of Alzheimer's disease. These conditions may provide a valuable clue to the pathogenesis of the commoner, multigenic late-onset forms; Alzheimer's disease is a particularly good example. However, in the case of the rare familial forms of type II diabetes, though some of the genes involved have been determined, they do not seem to have much relevance to the common multigenic variety of the disease.

It is simply too early to determine how successful this venture will be. It seems likely that some of the genes involved in susceptibility to these complex disorders will be identified and more will be learnt about their pathophysiology, knowledge which may lead to the generation of more effective forms of treatment. But given the extraordinary complexity of these conditions, it seems unlikely that information will be obtained which will be of genuine use for predictive genetics or focused public health for a very long time.

What Does the Future Hold?

The next few decades are likely to be among the most exciting in biology so far. The post-genomic period, during which attempts will be made to determine the function of many of our 30,000 genes and how they interact with one another and with the environment to make us what we are, is surely the most exciting venture that biological research has embarked on. It promises to help us to answer such fundamental questions as the working of the human brain, the mechanisms of behaviour, how the complex process of development is controlled, and the evolutionary history of human populations.

It is very difficult at the moment to see how far technological fall-out from these endeavours will impinge on medical research and practice. There seems little doubt that the pace at which this will happen has been over-exaggerated and that a more realistic view has to be taken. The tools of post-genomics have great potential for application to medical research and there seems little doubt that over the next few decades we will learn a great deal about the pathogenesis of disease and of the way in which pathogens invade their hosts and how they evade their hosts defence mechanisms. Equally, it is very likely that there will be a fall-out of valuable leads for the pharmaceutical industry and that there will be some slow progress towards somatic gene therapy, at least for monogenic diseases and for short-term gains in the management of common acquired disorders.

It may be that, ultimately, we shall see the day when there is widespread genetic screening and that every individual has their genome searched at birth for monogenic disease or susceptibility to the common disorders of middle life. And it is even possible that pharmacogenomics will start to play a role in therapeutic decision making. But given the complexities of even the simplest genetic disease, it may be a long time before any of these developments become a regular part of day to day clinical practice. For this reason it is very important that a balance is established between support for more conventional and well-tried approaches to clinical and epidemiological research and research into genomics as applied to medical practice. It is equally important that, if the fruits of the genome endeavour for medical care are to be fully realised, a closer partnership is struck up between clinicians who understand the complexities of disease and basic scientists who offer the tools for the dissection of these complexities. Unless there is increasing integration between the different branches of the biomedical sciences, much of the naive hyperbole which is bedevilling this field will continue. There is a danger in this because societies that are continually told that a new era of health care is just around the corner, yet the corner is never rounded, become disillusioned and the field loses credibility. Given its long-term potential, it would be a tragedy were this to happen.

Further Reading

Collins FS and McKusick VA (2001). Implications of the Human Genome Project for medical science. *J. Amer. Med. Ass.* **285**, 540-544.

Jiminez-Sanchez G, Childs B and Valle D (2001). Human disease genes. *Nature* **409**, 853-855.

Kaji EH and Leiden JB (2001). Gene and stem cell therapies. *J. Amer. Med. Ass.* **285**, 545-550.

Livingstone DM and Shivdasani R (2001). Towards mechanism-based cancer care. *J. Amer. Med. Ass.* **285**, 588-593.

Singer PA and Daar AS (2001). Harnessing genomics and biotechnology to improve global health equity. *Science* **294**, 87-89.

Weatherall DJ (1999). From genotype to phenotype: genetics and medical practice in the new millenium. *Phil. Trans. Roy. Soc. Lond.* B **354**, 1995-2012.

World   Health   Report   2000.   World   Health   Organization,   Geneva.

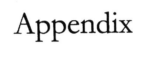

Appendix

JOURNAL OF THE ROYAL HORTICULTURAL SOCIETY
(1901) 25:54-61

PROBLEMS OF HEREDITY AS A SUBJECT FOR
HORTICULTURAL INVESTIGATION.

By Mr. W. BATESON, M.A., F.R.S., Fellow of St. John's College, Cambridge.

[May 8, 1900.]

AN exact determination of the laws of heredity will probably work more change in man's outlook on the world, and in his power over nature, than any other advance in natural knowledge that can be foreseen.

There is no doubt whatever that these laws can be determined. In comparison with the labour that has been needed for other great discoveries it is even likely that the necessary effort will be small. It is rather remarkable that while in other branches of physiology such great progress has of late been made, our knowledge of the phenomena of heredity has increased but little; though that these phenomena constitute the basis of all evolutionary science and the very central problem of natural history is admitted by all. Nor is this due to the special difficulty of such inquiries so much as to general neglect of the subject.

It is in the hope of inducing others to pursue these lines of investigation that I take the problems of heredity as the subject of this lecture to the Royal Horticultural Society.

No one has better opportunities of pursuing such work than horticulturists. They are daily witnesses of the phenomena of heredity. Their success depends also largely on a knowledge of its laws, and obviously every increase in that knowledge is of direct and special importance to them.

The want of systematic study of heredity is due chiefly to misapprehension. It is supposed that such work requires a lifetime. But though for adequate study of the complex phenomena of inheritance long periods of time must be necessary, yet in our present state of deep ignorance almost of the outline of the facts, observations carefully planned and faithfully carried out for even a few years may produce results of great value. In fact, by far the most appreciable and definite additions to our knowledge of these matters have been thus obtained.

There is besides some misapprehension as to the kind of knowledge which is especially wanted at this time, and as to the modes by which we may expect to obtain it. The present paper is written in the hope that it may in some degree help to clear the ground of these difficulties by a preliminary consideration of the question, How far have we got towards an exact knowledge of heredity, and how can we get further?

Now this is pre-eminently a subject in which we must distinguish what we *can* do from what we want to do. We *want* to know the whole truth of the matter; we want to know the physical basis, the inward and essential nature, "the causes," as they are sometimes called, of heredity. We want also to know the laws which the outward and visible phenomena obey.

Let us recognise from the outset that as to the essential nature of these phenomena we still know absolutely nothing. We have no glimmering of an idea as to what constitutes the essential process by which the likeness of the parent is transmitted to the offspring. We can study the processes of fertilisation and development in the finest detail which the microscope manifests to us, and we may fairly say that we have now a thorough grasp of the visible phenomena; but of the nature of the physical basis of heredity we have no conception at all. No one has yet any suggestion, working hypothesis, or mental picture that has thus far helped in the slightest degree to penetrate beyond what we see. The process is as utterly mysterious to us as a flash of lightning is to a savage. We do not know what is the essential agent in the transmission of parental characters, not even whether it is a material agent or not. Not only is our ignorance complete, but no one has the remotest idea how to set to work on that part of the problem. We are in the state in which the students of physical science were in the period when it was open to anyone to believe that heat was a material substance or not, as he chose.

But apart from any conception of the essential modes of transmission of characters, we can study the outward facts of the transmission. Here, if our knowledge is still very vague, we are at least beginning to see how we ought to go to work. Formerly naturalists were content with the collection of numbers of isolated instances of transmission—more especially, striking and peculiar cases—the sudden appearance of highly prepotent forms, and the like. We are now passing out of that stage. It is not that the interest of particular cases has in any way diminished— for such records will always have their value—but it has become likely that general expressions will be found capable of sufficiently wide application to be justly called "laws" of heredity. That this is so is due almost entirely to the work of Mr. F. Galton, to whom we are indebted for the first systematic attempt to enunciate such a law.

All laws of heredity so far propounded are of a statistical character and have been obtained by statistical methods. If we consider for a moment what is actually meant by a "law of heredity" we shall see at once why these investigations must follow statistical methods. For a "law" of heredity is simply an attempt to declare the course of heredity under given conditions. But if we attempt to predicate the course of heredity we have to deal with conditions and groups of causes wholly unknown to us, whose presence we cannot recognise, and whose magnitude we cannot estimate in any particular case. The course of heredity in particular cases therefore cannot be foreseen.

Of the many factors which determine the degree to which a given character shall be present in a given individual only one is known to us, namely, the degree to which that character is present in the parents. It is common knowledge that there is not that close correspondence between parent and offspring which would result were this factor the only one operating; but that, on the contrary, the resemblance between the two is only a general one.

In dealing with phenomena of this class the study of single instances reveals no regularity. It is only by collection of facts in great numbers, and by statistical treatment of the mass, that any order or law can be perceived. In the case of a chemical reaction, for instance, by suitable means the conditions can be accurately reproduced, so that in every individual case we can predict with certainty that the same result will occur. But with heredity it is somewhat as it is in the case of the rainfall. No one can say how much rain will fall to-morrow in a given place, but we can predict with moderate accuracy how much will fall next year, and for a period of years a prediction can be made which accords very closely with the truth.

Similar predictions can from statistical data be made as to the duration of life and a great variety of events the conditioning causes of which are very imperfectly understood. It is predictions of this kind that the study of heredity is beginning to make possible, and in that sense laws of heredity can be perceived.

We are as far as ever from knowing why some characters are transmitted, while others are not; nor can anyone yet foretell which individual parent will transmit characters to the offspring, and which will not; nevertheless the progress made is distinct.

As yet investigations of this kind have been made in only a few instances, the most notable being those of Galton on human stature, and on the transmission of colours in Basset hounds. In each of these cases he has shown that the expectation of inheritance is such that a simple arithmetical rule is approximately followed. The rule thus arrived at is that of the whole heritage of the offspring the two parents together on an average contribute one half, the four grandparents one quarter, the eight great-grandparents one eighth, and so on, the remainder being contributed by the remoter ancestors.

Such a law is obviously of practical importance. In any case to which it applies we ought thus to be able to predict the degree with which the purity of a strain may be increased by selection in each successive generation.

To take a perhaps impossibly crude example, if a seedling show any particular character which it is desired to fix, on the assumption that successive self-fertilisations are possible, according to Galton's law the expectation of purity should be in the first generation of self-fertilisation 1 in 2, in the second generation 3 in 4, in the third 7 in 8, and so on.

But already many cases are known to which the rule in the simple form will not apply. Galton points out that it takes no account of individual prepotencies. There are, besides, numerous cases in which on crossing two varieties the character of one variety is almost always transmitted to the first generation. Examples of these will be familiar to those who have experience in such matters. The offspring of the Polled Angus cow and the Shorthorn bull is almost invariably polled. Seedlings raised by crossing *Atropa belladonna* with the yellow-fruited variety have without exception the blackish-purple fruits of the type. In several hairy species when a cross with a glabrous variety is made, the first cross-bred generation is altogether hairy.

Still more numerous are examples in which the characters of one variety very largely, though not exclusively, predominate in the offspring.

These large classes of exceptions—to go no further—indicate that, as we might in any case expect, the principle is not of universal application, and will need various modifications if it is to be extended to more complex cases of inheritance of varietal characters. No more useful work can be imagined than a systematic determination of the precise "law of heredity" in numbers of particular cases.

Until lately the work which Galton accomplished stood almost alone in this field, but quite recently remarkable additions to our knowledge of these questions have been made. In the present year Professor de Vries published a brief account[*] of experiments which he has for several years been carrying on, giving results of the highest value.

The description is very short, and there are several points as to which more precise information is necessary both as to details of procedure and as to statement

---

[*] *Comptes Rendus*, March 26, 1900, and *Ber.d.Deutsch.Bot.Ges.*, xviii.1900, p.83.

of results.† Nevertheless it is impossible to doubt that the work as a whole constitutes a marked step forward, and the full publication which is promised will be awaited with great interest.

The work relates to the course of heredity in cases where definite varieties differing from each other in some *one* definite character are crossed together. The cases are all examples of discontinuous variation: that is to say, cases in which actual intermediates between the parent forms are not usually produced on crossing. It is shown that the subsequent posterity obtained by self-fertilising these cross-breds or hybrids break up into the original parent forms according to fixed numerical rule.

Professor de Vries begins by reference to a remarkable memoir by Gregor Mendel,‡ giving the results of his experiments in crossing varieties of *Pisum sativum*. These experiments of Mendel's were carried out on a large scale, his account of them is excellent and complete, and the principles which he was able to deduce from them will certainly play a conspicuous part in all future discussions of evolutionary problems. It is not a little remarkable that Mendel's work should have escaped notice, and been so long forgotten.

For the purposes of his experiments Mendel selected seven pairs of characters as follows:–

1. Shape of ripe seed, whether round, or angular and wrinkled.

2. Colour of "endosperm" (cotyledons), whether some shade of yellow, or a more or less intense green.

3. Colour of the seed-skin, whether various shades of grey and grey-brown, or white.

4. Shape of seed-pod, whether simply inflated, or deeply constricted between the seeds.

5. Colour of unripe pod, whether a shade of green, or bright yellow.

6. Shape of inflorescence, whether the flowers are arranged along on axis, or are terminal and more or less umbellate.

7. Length of peduncle, whether about 6 or 7 inches long, or about ¾ to 1½ inch.

Large numbers of crosses were made between Peas differing in respect of each of these pairs of characters. It was found that in each case the offspring of the cross exhibited the character of one of the parents in almost undiminished intensity, and intermediates which could not be at once referred to one or other of the parental forms were not found.

In the case of each pair of characters there is thus one which in the first cross prevails to the exclusion of the other. This prevailing character Mendel calls the *dominant* character, the other being the *recessive* character.*

That the existence of such "dominant" and "recessive" characters is a frequent phenomenon in cross-breeding, is well known to all who have attended to these subjects.

---

† For example, I do not understand in what sense de Vries considers that Mendel's law can be supposed to apply even to all "monohybrids," for numerous cases are known in which no such rule is obeyed.

‡ 'Versuche üb. Pflanzenhybriden' in the *Verh. d. Naturf. Ver. Brünn*, iv. 1865.

* Note that by these useful terms the complications involved in the use of the expression "prepotent" are avoided.

By self-fertilising the cross-breds Mendel next raised another generation. In this generation were individuals which showed the dominant character, but also individuals which preserved the recessive character. This fact also is known in a good many instances. But Mendel discovered that in this generation the numerical proportion of dominants to recessives is approximately constant, being in fact *as three to one*. With very considerable regularity these numbers were approached in the case of each of his pairs of characters.

There are thus in the first generation raised from the cross-breds 75 per cent. dominants and 25 per cent. recessives.

These plants were again self-fertilised, and the offspring of each plant separately sown. It next appeared that the offspring of the recessives *remained pure recessive,* and in subsequent generations never reverted to the dominant again.

But when the seeds obtained by self-fertilising the dominants were sown it was found that some of the dominants gave rise to pure dominants, while others had a mixed offspring, composed partly of recessives, partly of dominants. Here also it was found that the average numerical proportions were constant, those with pure dominant offspring being to those with mixed offspring as one to two. Hence it is seen that the 75 per cent. dominants really are not all alike, but consist of twenty-five which are pure dominants and fifty which are really cross-breds, though, like the cross-breds raised by crossing the two varieties, they only exhibit the dominant character.

To resume, then, it was found that by self-fertilising the original cross-breds the same proportion was always approached, namely—

25 dominants, 50 cross-breds, 25 recessives, or 1D : 2DR : 1R.

Like the pure recessives, the pure dominants are thenceforth pure, and only give rise to dominants in all succeeding generations.

On the contrary the fifty cross-breds, as stated above, have mixed offspring. But these, again, in their numerical proportions, follow the same law, namely, that there are three dominants to one recessive. The recessives are pure like those of the last generation, but the dominants can, by further self-fertilisation and cultivation of the seeds produced, be shown to be made up of pure dominants and cross-breds in the same proportion of one dominant to two cross-breds.

The process of breaking up into the parent forms is thus continued in each successive generation, the same numerical law being followed so far as has yet been observed.

Mendel made further experiments with *Pisum sativum*, crossing pairs of varieties which differed from each other in two characters, and the results, though necessarily much more complex, showed that the law exhibited in the simpler case of pairs differing in respect of one character operated here also.

Professor de Vries has worked at the same problem in some dozen species belonging to several genera, using pairs of varieties characterised by a great number of characters: for instance, colour of flowers, stems, or fruits, hairiness, length of style, and so forth. He states that in all these cases Mendel's law is followed.

The numbers with which Mendel worked, though large, were not large enough to give really smooth results; but with a few rather marked exceptions the observations are remarkably consistent, and the approximation to the numbers demanded by the law is greatest in those cases where the largest numbers were used. When we consider, besides, that Tschermak and Correns announce definite confirmation in the case of *Pisum*, and de Vries adds the evidence of his long series of observations on other species and orders, there can be no doubt that Mendel's law is a substantial reality; though whether some of the cases that depart most widely from it can be brought within the terms of the same principle or not, can only be decided by further experiments.

One may naturally ask, How can these results be brought into harmony with the facts of hybridisation as hitherto known; and, if all this is true, how is it that others who have so long studied the phenomena of hybridisation have not long ago perceived this law? The answer to this question is given by Mendel at some length, and it is, I think, satisfactory. He admits from the first that there are undoubtedly cases of hybrids and cross-breds which maintain themselves pure and do not break up. Such examples are plainly outside the scope of his law. Next he points out, what to anyone who has rightly comprehended the nature of discontinuity in variation is well known, that the variations in each character must be separately regarded. In most experiments in crossing, forms are taken which differ from each other in a multitude of characters—some continuous, others discontinuous, some capable of blending with their contraries, while others are not. The observer on attempting to perceive any regularity is confused by the complications thus introduced. Mendel's law, as he fairly says, could only appear in such cases by the use of overwhelming numbers, which are beyond the possibilities of practical experiment.

Both these answers should be acceptable to those who have studied the facts of variation and have appreciated the nature of Species in the light of those facts. That different species should follow different laws, and that the same law should not apply to all characters alike, is exactly what we have every right to expect. It will also be remembered that the principle is only declared to apply to discontinuous characters. As stated also it can only be true where reciprocal crossings lead to the same result. Moreover, it can only be tested when there is no sensible diminution in fertility on crossing.

Upon the appearance of de Vries' papers announcing the "rediscovery" and confirmation of Mendel's law and its extension to a great number of cases two other observers came forward and independently describe series of experiments fully confirming Mendel's work. Of these papers the first is that of Correns,[*] who repeated Mendel's original experiment with Peas having seeds of different colours. The second is a long and very valuable memoir of Tschermak,[†] which gives an account of elaborate researches into the results of crossing a number of varieties of *Pisum sativum*. These experiments were in many cases carried out on a large scale, and prove the main fact enunciated by Mendel beyond any possibility of contradiction. Both Correns (in regard to Maize) and Tschermak in the case of *P. sativum* have obtained further proof that Mendel's law holds as well in the case of varieties differing from each other in two characters, one of each being dominant, though of course a more complicated expression is needed in such cases.[‡]

That we are in the presence of a new principle of the highest importance is, I think, manifest. To what further conclusions it may lead us cannot yet be foretold.

---

[*] *Ber. deut. Bot. Ges.*, 1900, xviii. p. 158.

[†] *Zeitschr. f. d. landw. Versuchswesen in Oesterr.*, 1900, iii. p. 465.

[‡] Tschermak's investigations were besides directed to a re-examination of the question of the absence of beneficial results on cross-fertilising *P. sativum*, a subject already much investigated by Darwin, and upon this matter also important further evidence is given in great detail.

But both Mendel and the authors who have followed him lay stress on one conclusion, which will at once suggest itself to anyone who reflects on the facts. For it will be seen that the results are such as we might expect if it is imagined that the cross-bred plant produced pollen grains and ovules, each of which bears only *one* of the alternative varietal characters and not both. If this were so, and if on the average the same number of pollen grains and ovules partook of each of the two characters, it is clear that on a random assortment of pollen grain and ovules Mendel's law would be obeyed. For 25 per cent. of "dominant" pollen grains would unite with 25 per cent. "dominant" ovules; 25 per cent. "recessive" pollen grains would similarly unite with 25 per cent. "recessive" ovules; while the remaining 50 per cent. of each kind would unite together. It is this consideration which leads both de Vries and Mendel to assert that these facts of crossing prove that each ovule and each pollen grain is pure in respect of each character to which the law applies. It is highly desirable that varieties differing in the form of their pollen should be made the subject of these experiments, for it is quite possible that in such a case strong confirmation of this deduction might be obtained.

As an objection to this deduction, however, it is to be noted that though true intermediates did not occur, yet the degrees in which the characters appeared did vary in degree, and it is not easy to see how the hypothesis of perfect purity in the reproductive cells can be supported in such cases. Be this, however, as it may, there is no doubt we are beginning to get new lights of a most valuable kind on the nature of heredity and the laws which it obeys. It is to be hoped that these indications will be at once followed up by independent workers. Enough has been said to show how necessary it is that the subjects of experiment should be chosen in such a way as to bring the laws of heredity to a real test. For this purpose the first essential is that the differentiating characters should be few, and that all avoidable complications should be got rid of. Each experiment should be reduced to its simplest possible limits. The results obtained by Galton, and also the new ones especially detailed in this paper, have each been reached by restricting the range of observation to one character or group of characters, and there is every hope that by similar treatment our knowledge of heredity may be rapidly extended.

[Note.–Since the above was printed further papers on Mendel's Law have appeared, namely, de Vries, *Rev. génér. Bot.,* 1900, p. 257; Correns, *Bot. Ztg.,* 1900, p. 229; and *Bot. Cblt.,* lxxiv., p. 97, containing new matter of importance. Prof. de Vries kindly writes to me that in asserting the general applicability of Mendel's Law to "monohybrids" (crosses between parents differing in respect of *one* character only), he intends to include cases of discontinuous varieties only, and he does not mean to refer to continuous varieties at all. October 31, 1900.]

T - #0446 - 071024 - C180 - 234/156/8 - PB - 9780367394462 - Gloss Lamination